SCIENTIFIC EXPLOITS:
The Glorious, the Humorous, and the Serious

Copyright 2010 Austin P. Torney
austintorn@aol.com

I0475518

ON THE ORIGIN,
WHO DESIRED THAT ON ITS TOMB
SHOULD BE INSCRIBED--

"Here lieth One whose name was writ on water."

The 'false' and melted vacuum was liquid energy—
Unstructured, unordered, and going nowhere,
But, then, inexplicably, it 'fell',
As from a kind of 'shelf'…

…Whirling, twirling and swirling inward
Until there was no more inward left…

It 'thought' that its future could never be,
That its quality was but written
On the water and the wind
With a feathery quill
Whose ink was the smoke and fog
Of a shimmering dream.

Then it died… like the Phoenix.

And thus it crystalized, frozen,
Into our structured 'true' vacuum…

For, ere the breath that could erase it blew,
Death, in remorse for that fell slaughter,
Death, the immortalizing winter, flew
Athwart the flowing stream—
And Time's printless torrent grew
A scroll of crystal,
Blazoning the name
Of 'The Universe'!

THE NEAR DECLINE OF PHYSICS
DUE TO ITS UNDRESSED TERMS

The quarks, those constituents of the orgy
Playfully bound within the nucleons' chamber
Are named *up, down, strange, charm, bottom and top,*
The last two once being called *beauty,* and *truth;*

However, when just one of a type was contained
It became referred to, say, as a naked beauty,
And thus nude tops and bottoms their charms revealed—
To ever be in closeness binding, and bonding,

So, they even tried just *u, d, s, c, b,* and *t*
To prevent some ultimate collapse of physics,
But the truth of the flavors beneath the veils
Remained as the sheerest vision preferred.

So, we have these vibrant dancing ladies:
The naked heavyweight top, charming up,
And, down, the strange beauty of the raw truth,
With a bare bottom just around and behind.

They gyrate, spinning their charms, twirling,
In the universal dance of stunning motion,
The polarity sometimes reversed,
Whirling, their bottoms up and tops down.

And then there are Eden's many colors,
In this flower garden filled with flavors,
Such as red bottom beauties, blue tops,
And magenta undulations unstopped.

Gluons are the bees of the flower beds,
Carrying pollen back and forth to bond
The many relationships that make
This loved world go 'round as reality.

Eyed in views that probe the fundamental,
Quarks strangely swirl in and out of sight,
Pulsing, throbbing with elemental delight,
In and out—the love-made life of eternity.

These attractions in the altogether denuded
In the buff became the strong force, manifest,
That the mother-nature-naked terms exposed
To denote the stark beauty of truth uncovered.

"THREE QUARKS FOR MUSTER MARK"

Naked quarks would really love to go wild and dance,
But there's only a finite amount of energy and chance;
So, they would spiral out of control, having quite a blast!
Such, they have been confined within the proton—to last.

They're made bottoms-up;
Can we see them tops-down, a go-go?

No, for the quantum censor protects the charm show,
Their strange beauty and flavor bound up and down,
As the proton is much immune to disturbance around.

TRUTH OR HAPPENSTANCE?

John Allen Paulos said that
"In reality, the most astonishingly
Incredible coincidence imaginable
Would be the complete absence of coincidence."

Einstein was bothered by the fact
That gravity and inertia balance out exactly,
Allowing a cannon ball to fall
At the same rate as a marble.

Was it a happenstance or a truth?

He concluded that gravity and inertia
Are aspects of a bigger picture—
The curvature of space-time.

THAT BLANKETY BLANK

Euclid and Pythagorus never even thought of it,
Perhaps not needing it for geometry;
So it was and wasn't 'Greek' to them.

Aristotle was deathly afraid of it.

Even the word 'naughty' came from it.

'0' had a chilly reception everywhere,
It's rounded symbol enclosing nothing,
As if it could be captured,
But 'nothing' never changed,
Being the same even if you took it away.

Humans stumbled on zero & nothing by accident,
And recoiled in horror, fearing it, reviling it,
And sometimes even banning it outright
As some kind of evil influence.

After many centuries, it seemed to be tamed,
Put in its place as a simple little placeholder.

Then the beast reared its ugly head for real,
Misbehaving like a monster right and left:

It brought instant death by multiplication,
And wrought total absurdity through division,
Yet halting our expensive computers.

It exploded into the ambiguous fog of infinity;
It ran away from us in calculus,

Sliding us down the slippery slope
Of closing in on it but never reaching it.

It spawned ghosts such as negative numbers,
Imaginaries, and those ephemeral infinitesimals.

Both the genie and the genius
Had been let out of the bottle,
And the goose egg still
Confounds and confuses,
No one knowing zilch about it,
Creating paradoxes left and right.

NOTES ON NOTHING

If something could be made from Nothing,
Anything could and would
Spring out anywhere, anytime.

"Nature abhors a vacuum"

— Aristotle

"Absolute Nothing cannot be;
Therefore there must be something"

— Austin

TALKING TOO MUCH

On and on we say of What paved the way,
Then even tell the nature of such Theity,
And on and on we presume further upon,
Joining that group called 'On and on anon'.

STAR VOYAGERS

Some quarks we'll take aboard the final ark,
On that penultimate day that all goes dark
As the Red Giant envelopes Terra with fire,
Then rebuild the Earth closer to our desire.

HOW THE ALLIES WON WORLD WAR II

Warner Heisenberg, the head of
The German Nuclear Weapons Effort,
Was full of the uncertainty
That he had discovered in physics.

Heisenberg was entangled with his old mentor,
The Danish physicist Neils Bohr,
They being old friends, like father and son.

They were also supposed to be enemies,
For Germany occupied Denmark.

Together they had created a physics
Of deep truth and beauty,
For beauty was the expression of truth.

They also made possible the physics
To destroy large cities,
Even the entire world.

In 1941, Heisenberg went to see Bohr,
The 'father of quantum mechanics',
In Copenhagen, Denmark;
But we don't know what they discussed;
Yet, Germany failed to complete its work
To build an atomic bomb.

Did Heisenberg deliberately withhold
Information from the Nazis?

Did this consummate mathematician
Neglect to perform an obvious calculation?

Did he, with Bohr,
Form a complimentery pair,
Joining their views
Of the political position versus its velocity
To form a complete picture of reality?

Did a man's heart turn the tide of War?

STAR STUFF

How could we ever know
The composition of a star?

It's not like we could go there
To collect a sample.

"Impossible," it was thought.

Then starlight shadows were found that
Spelled out a complete list of the ingredients—
A quantum mechanical bar code of its elements.

ASTROLOGICAL NON-SCIENCE

The gravitational pull of Venus and Mars
On someone at birth would be overpowered
Even by the gravitational pull
Of the obstetrician walking
Around the delivery table.

THE IRRATIONAL

Indefiniteness didn't sit well with Pythagorus,
Ever concerned with the perfection of numbers.

You can divide the circumference of a circle
With its radius but you cannot write the result
As a fraction or a ratio, for pi just keeps on going.

So, he pledged his disciples to secrecy,
For the ancient Pythagoreans
Had developed an entire religion
Based on the rationality of numbers.

Yet, one rebel vowed to let the word out;
However, he mysteriously drowned at sea.

HALLEY, NEWTON, AND HOOKE

Halley was a sea captain, a cartographer, a professor
Of geometry, a deputy of the Royal Mint, an astronomer,
And the inventor of the deep-sea diving bell,
And wrote some on magnetism, tides,
Planet motions, and fondly on opium.

He invented the weather map and actuarial table ages,
Even proposed methods to work out the Earth's old age,
Its distance from the sun, even how to keep fresh fish,
But one thing he didn't do was to discover Halley' comet,
For he merely noted that it was yet another return of it.

He made a wager with Robert Hooke, the cell describer,
And with the great and stately Christopher Wren:
They bet upon why the planets' orbit were ellipses.

Hooke, a known credit-taker,
Claimed he'd solved the problem,
But had to conceal it
So that others could yet know the satisfaction.
Well, Halley became consumed with finding the answer,
So he called upon the Lucasian Mathematics Professor.

Isaac Newton was indeed brilliant beyond measure,
But was solitary, joyless, paranoid—no pleasure.

Once he had inserted a needle in his eye & poked around,
Far inserting the bodkin between the eye and the bone.
Another time, he'd stared at the sun for so very long
That he had to spend many days in a darkened room.

Frustrated by mathematics, Isaac invented the calculus,
And then for twenty-seven years kept it hidden from us.
Likewise, he did the same with the understanding of light
And spectroscopy, keeping it for thirty years in the dark.

For Newton,
Science was but a partial part
Of his life's routes,

For much of his time
Was given to alchemy
And religious pursuits.

He was wholeheartedly devoted
To the religion of Arianism,
Whose main tenet was
That there could be no Holy Trinity.

Ironically, he worked as a Professor at Trinity College,
Although the only one there who was not Anglican.
He also spent an inordinate amount of time studying
The floor plan of the lost temple of Solomon the King,
Even learning Hebrew, the better to scan the texts.

Another single minded quest was
To turn base metals Into precious ones,
His papers revealing this preoccupation
Over optics and planetary motions and such mentations.

Well, Halley asked Newton what the curve would be
If the planets' attraction toward the sun was
The reciprocal to the square of their distance from it.
Newton promptly answered, of course, an "ellipse".

Not finding his calculations of it
Newton not only rewrote it,
But retired for two years to produce his master work,
The Philosophiae Naturalis Principia Mathematica.

To Halley's horror,
Newton refused to release the crucial 3rd volume,
Without which the first two would make little sense.
There had been a dispute between Newton and Hooke
Over the priority of the inverse square law in the book.

That solved by Halley's diplomacy, the Royal Society
Had pulled out from the publication, failing financially,
For, the year before, there had been a very costly flop
Called *The History of Fishes;* so, Halley himself popped
The funds for the publication out of his own pocket.

Newton contributed nothing,
As usual, and, to make matters worse,
Halley had just taken a position as the society's clerk,
They failing to pay the promised 50 pounds to his purse,
Paying him only with very many copies of
The History of Fishes!

SEEDLESS

Oh, so many lofty words for God's sake,
But none of these do any Creator make.
The speaking mammals hide not their pride
That says of where the higher soul must reside.

But the only thing we can ever know
Is that we are starring in this reality show.
"I must know every answer," says the mind,
"Of the makings before our humankind."

But the mind could not answer in kind
Of the ancient secret left so far behind,
But, it wants to know, for that's its purpose,
So it ever fills in the blanks before us.

Then it comes to believe the non-sense,
The only way it can try to make sense.
Go along with its imaginings, thoughtlessly?
Or give the mind a rest from sowing aimlessly?

None can know God's truth with any proof
Of this reality show our mind to soothe.
This is not a restriction, since it ever frees
One to live as sensibly as they please.

Who is to punish you for not knowing
But that you are in this world growing,
Having become here willy-nilly going?
As life's rose, outspread your fragrance blowing.

COLOR SYMBOLS

In the netherworld, I learned the lore and
Legends of the colors, of their uses
In nature and emotions, the whatfor
Of their light's glowing activity:

All color variants, quite numberless,
Are made from the three primaries, no less;
Namely: red, yellow, and blue—often backed
By colorless white tinges or shades of black.

From just these three essential hues derives
All of heaven's prismatic radiance,
Myriad colors of floral brilliance,
And technicolors that come so alive.

The offspring of married red and yellow
Is the secondary, orange, a bright fellow;
Its sibling, of blue and yellow, is green,
With, of course, some gradation in between.

Saintly brother purple, twixt reds and blues,
Completes the second generation hues.
Next to arrive, lime-green, is a grandchild,
As are all the tertiary colors wild—

They're crimson, magenta, maroon, scarlet,
Amber, auburn, salmon, ocher, russet,
Mauve, taupe, fuchsia, cherry, cerise, umber,
Teal, emerald, and vermilion others.

Strangely enough, all the color-pairs
That symbolize seasons and festive fairs
As they're found naturally in nature's ways,
Do contrast on the color wheel, crossways:

Direct opposites on the color wheel,
Sky-blue and leafy-orange, represent fall,
For they are autumn's contrasting colors
That quite up for its lack of flowers.

As with crocus, spring's floral colors yet
Remain yellow primrose, purple violet—
The sensual sun, as it were, warming
The virginal earth, with love, into spring.

The Christmas Holiday Season is "seen"
In its opposing hues of red and green—
As in Holly, berry-red, ever-green,
Or in Poinsettias' red flush, leaf of green.

We're out of diametric color sets,
So, which for summer? It must then contain
The entire spectrum, as these the sunset
And the rainbow express, in shine and rain.

Since winter's snow hides all things out of sight,
Its colors are hidden inside white—and night,
The cold season's symbols, for they conceal
All of spring and summer's bright floral feel.

For that as different as day and night,
We have the twin-opposites: black and white,
For the day-clock first became dark and light
When twin-gods split day & night, wrong & right.

Heaven's splendor, white, for purity, bless,
Holds all the colors of prismatic light,
But the symbol of the Prince of Darkness,
Black, removes all the colors from our sight.

So then, it is proved that, in both nature
And in the color wheel, opposites attract
And complement in their contrast—to procure
Both real and symbolic color contracts.

Next, we'll turn to the colors lone, to see
The whatfor of their light's activity,
But first, let's ask, Are there any missing hues,
Unknown, hidden in rainbows, or not used?

Hidden colors? No, for I see how red goes
To orange, graduating through the rainbow
Into yellow and on through green, to let
Blue into indigo to become violet.

Perhaps, between green and blue, lies some new
Tincture, unique enough to be it's own hue,
But, alas, those turquoise waves everyday,
In tropic seas, wash that theory away.

Yet, there may be some new colors that lie
Before or beyond the spectrum and the eye,
Like infrared or ultraviolet,
Or gold, which only the fairies can see.

But what of clear, white, silver, gray, or black?
Well, they're not true colors, for, either they lack
All color (black, clear) or hide all hues (white)
Or are mixtures (gray, silver): black-white.

But wait, there is a well-known color,
One quite common in both dress and nature,
That cannot be found in the rainbow—
Give up? It's brown—and has nowhere to go!

Brown is the color of death, like the leaves
That crumble dry and lifeless when earth grieves,
Which is why the faeries won't let it show
In their magically spectral rainbow.

But, alas, brown's new hue is not to last,
For brown's no more than red, yellow, and black.
So, onward we move: What do colors mean?
What's nature's physiological scheme?

When we see red, we see danger: Stop! Blood!
Metabolism rises, adrenaline floods—
And, so, restaurants use red tablecloths
To increase both the appetite and the cost.

Yellow, the quickest color we can see,
Means caution, as with black on a bee,
But yellow's bright and cheerful, too, and lends
Light to small and sunless rooms like kitchens.

Healthful orange is the common man's color;
So, to make the expensive look cheaper,
Such as with a hotel, they paint it orange,
And put some shiny polish on the door hinge.

Blue invigorates, and, therefore, provides
Extra strength and power, so blue's on our side
When the home team's locker room is painted
In its hue (visitor's was pink—they fainted).

Blue, as was said, is good, except on food,
For few foods are blue; so, in diet mood,
Put a blue light in your kitchen—and lose
Weight avoiding repulsive looking food.

Pink (red tinted with white) debilitates,
Sapping strength and temper, so, that is why
It's used in prison cells and locker rooms,
For it calms the most violent inmates.

What of purple? Well, it's mournful, but, too,
It's stately, regal, and virginal, new.
Of green, though it's seldom worn, none complain;
And use it in their carpets to stay sane.

The stars are not just white, they scintillate:
Sirius is blue, its companion green;
Betelgeuse, red; many, like Sol, yellow;
Arcturus, orange—all jewels constellate.

Well, as colors go, so, then, do we, see:
Hues are just differing wavelengths of light
That the brain interprets, in its own right,
For some natural colored necessity.

May I chance upon a land of strange rainbows
Of elfin-hued flowers: red delphiniums,
Black tulips, orange fuchsias, white marigolds,
Bronze grass, and the legendary blue rose.

SIMPLICITY

Occam sharpened his razor,
To a one-dimensional line,
Then cut his beard into strings.

They sprung from the depths,
Vibrating the songs of reality,
For which all composites sprang.

THE RAZOR

In the alphabet, Occam saw the unnecessary,
So he struck out 'j', 'q', 'x", and 'z',
Being rarities or duplicates,
And then even cut more,
Those being the vowels taking up space.

n th lphbt, ccm sw th nncssry,
S h strck t ", ", ", and ",
Bng rrts r dplcts,
nd thn vn ct mr,
Ths bng th vwls tkng p spc.

(But then one could only understand him almost)
(Bt thn n cld nly ndrstnd hm lmst)

CONCISE SIMPLICITY

Writers of few words,
Even the laws that writ reality,
Can often say more more with less.

NEW 9TH PLANET FOUND!

Poor Pluto's been banished to the underworld,
Charon rowing him to the land of the forgotten.
Schoolchildren petitioned for his return,
But he was voted off of the solar island.

Memory's crutch for the order of the planets,
Is now just "MVEMJSUN"—
Old Pluto tried so darn hard, its position
Now even closer to the sun than Neptune's.

Well, many have searched for quite a while for
The next planet without any success—
There have been hoaxes, theories, and some ghosts;
Yet, I have firm proof of another planet.

But, first, a review of some poor attempts:
'Vulcan' was spotted very close to the sun,
And 'existed' for about five days,
But now is relegated to the Star Trek World.

Another 'Vulcan', impossible to see,
Being 180 degrees away from Earth,
Behind the sun, was seen in the movie
"Journey to the Far Side of the Sun".

Could an asteroid like Eris be a planet?
Nope, 'tis not allowed, although all of the
Debris between Mars and Jupiter
Could have come from an unstable planet.

Nice try, but it's not out there anymore,
And any planets of other solar systems
Don't count, nor does Planet Hollywood
Or Daily Planet or any other restaurants.

Perhaps there's another planet way out,
Beyond; that may be so, but, no matter,
Though it may become the 10th planet, since
I have found the newest 9th, with no doubt.

The 9th planet does follow an orbit
Close to Earth's, ever falling toward the sun—
It is right under our nose: It's the moon!
But, wait, you say, it is Earth's satellite.

Our moon is unique in the solar system—
It's not captured by the Earth, but by the sun,
It's orbit being everywhere concave to Sol.
(Thanks to Isaac Asimov for proving this.)

Never does our moon fall away from the sun,
For it's attracted to it about twice as much
As it is to the Earth, although the moon and
The Earth do form a double planet system

That revolves about a common point that
Happens to be inside of the Earth.

SOUL JUMPING BROUGHT TO A CONCLUSION

There is a clear and powerful motive for us
To *want* to believe in a soul beyond dust to dust.

The soul is a potential lifeline to forever being,
A way to avoid the finality of death's knell ringing.

What a great thing if there were a part of 'I'
That could not die—a reward of pie in the sky!

So, we have a vested interest in soul theory—
And *this fact alone* is sufficient to account
For the belief in souls and an after-whence
In spite of a conspicuous lack of evidence.

Nor is the mind distinct from the living brain,
Upon which our memories and thoughts rain
And surface upon, a select few at a time, that
Neurologists can observe deep within our hat.

GRAVITY'S WELL

Gravity fell, from its fundamental throne,
Being a blend of matter and motion.
As with time, if we take away what's known,
Its attraction fades into oblivion.

It is already gone in my dream sleep
In which I float, fly, and hover at will;
But, upon awakening from the deep,
The super bed-gravity holds me still.

Instead of dieting, I live on the moon,
Playing golf, but the bunkers are so deep
I have to take some giant leaps until noon;
'Though I love the freedom of low-grav feet.

If there were none, life could really be tough,
Our stuff floating away—what losing brings;
What a mess, although it might help those fluffs
Mercilessly dominated by material things.

If gravity's of movement and matter,
Like time, it could be a new dimension,
So to speak, but may still need the other three,
Although it's just the right dose of tension.

We can conquer gravity's whole world round
By the mere lift of a little finger;
Yet we get hang-ups about our hang-downs,
And thoughts of what the hell it is still linger.

THE GREAT EXTINGUISHER

Our planet is very good at promoting life,
But it is much better at extinguishing it.

Of the billions upon billions of living things,
99.99% are no longer around here living.

THE JUNE 30, 1860 SHOWDOWN

Were we descended from some ape-like creatures?
A thousand people sat down to hear the lectures.

The Bishop of Oxford, Samuel Wilberforce, rose to speak,
And, while speaking, and into his flow, looked at Huxley,
And asked if he'd become attached to apes by way
Of his grandmother's or his grandfather's recent sway.

Huxley turned to his neighbor and whispered plans,
"The Lord has delivered him into my hands",
Then rose with a relish and said something, agape,
Of the nature "I'd rather claim kinship to an ape
Than to someone using his eminence to propound
Such unscientific twaddle in a serious scientific forum!"

This was an insult to the Bishop's office and his door,
So, the proceedings instantly turned into an uproar.

Someone ran around holding up a Bible, to exclaim
"The Book, the Book!" (Truly, we'll never be the same.)

Now, who was this guy holding up the Book?
It was none other than the pilot of the Beagle.

THE CONCEPTION OF NATURAL LAWS

You cannot fool Mother Nature; it is improper,
For thou shalt not fiddle with Mother Nature;
But, since Father Time outlives all who venture,
He can fool around with Mother Nature.

At some time during a long eternity,
His paternity begot her maternity;
They then gave birth to life's certainty.

ALCHEMY HAPPENS VIA RADIOACTIVITY
AND HOW OLD CAN THE EARTH BE?

Through E=MCC we see that vast energy reserves
Are bound up in small amounts of matter, preserved.

Henri Becquerel carelessly left a packet of uranium salts
On a wrapped photographic plate in his drawer vault.

Some time later, he was surprised to discover that
The salts had burned a 'light' impression into it.
The salts were emitting rays of some sort, curiously,
So, he turned the matter over to Marie Curie, literally.

Madam Curie and her new husband Pierre, with glee,
Noted that the rocks poured out great amounts of energy,
But they never diminished in size or changed in any way.
They were converting mass into energy very efficiently.

They also found polonium and radium, and a Nobel prize,
Along with Becquerel, in 1903, Einstein yet on the rise.

Radioactive elements decayed into other elements,
Noted Ernest Rutherford and colleague Fredrick Soddy;
One day you had an atom of uranium that "bled",
And the very next day you had an atom of lead.

It always took the same amount of days.
For half of the sample to decay,
And so this steady reliable rate of decay
Could be used in kind of a clocking way.

Tick-tock, how old was it?
More than 700 millions years worth!
This age was way more
Than anyone had given the Earth.
(5 billion would be closer to the answer.)

He lectured one day,
Taking out a piece of radioactive pitchblende,
Showing it to aging Kelvin,
But Kelvin rejected it to the end.

Dimitri Mendedeyev rejected it too,
As with everything new,
Ever storming out of labs
And lecture halls all over, too;
However, the 101st element
Was called mendelevium,
In his name meant,
And quite appropriately,
For it was a very unstable element.

Pierre Curie began to experience
Radiation sickness, getting weak,
But in 1906 he was fatally run over
By a carriage on a Paris street.

Marie worked on with much distinction,
But had an affair so indiscreet
That even the French were scandalized there,
And so she was never elected
To the Academy of Sciences,
Despite not just one,
But two Nobel prizes
(Physics, Chemistry).

Scientists yet thought that radioactivity was beneficial,
Putting thorium into toothpaste and laxatives as useful;
Eventually these products were banned, by 1938,
But for Madam, who'd died of leukemia,
It was much too late.

The radiation is so pernicious and long lasting
That even now her papers from the 1890's,
And even her cookbooks, are dangerous and toxic,
So, all her lab books must be kept in lead lined boxes.
(One must wear protective clothing to look at them.)

Marie Curie was a very attractive lady, very much aglow,
For my great ancestor in his old writings such told me so.
She radiated warmth unto him as a rainbow of sparks—
"Great balls of fire!" he remarked,
"They now glow in the dark!"

FROM MATTER TO US

The big bang, or materialization,
One of many, was prosperous for us,
For its constants allowed for life's basis
'Though it didn't have us in mind at all.

It arose from some unbreakable stuff,
Perhaps several such eternal things,
Or the same from previous contraction;
But not from nothing, for how could this be?

Now, if the big bang's material result
Was not favorable for our becoming,
Then we wouldn't be here to discuss it.
'Though auspicious, it guaranteed nothing.

Matter and antimatter formed of it,
In equal parts, most of it annihilating;
However, some black holes evaporated,
Leaving a fortunate amount of matter.

Matter's here that works as building blocks,
The strong force's stability opposed
To the weak force's dispersal through decay,
Plus electromagnetism's motion.

Lucky, not planned, all this gave us a chance,
As from the stars cameth our help and hope,
When they generated all the elements
That brewed a soup of fortuitous accidents.

Earth was a golden distance from the sun—
A large number of other planets unfit;
Earth's features evolved in a good proportion
To sustain the beginnings of early life.

Soil and bacteria generated oxygen;
Death chose the useful forms over the useless,
Kept track of by RNA-like structures
In life's cradle, though we had not yet appeared.

Our blind fated path was the further paved
When asteroids finished most of the species—
Far from a feature of intelligent design,
But, it opened the space that was needed.

Evolution sifted through the accidents,
As it directed the good from the bad;
We began from the fusing of chromosomes
That made us incompatible with "chimps".

From matter to us—to all our senses,
To our brains, our minds, and consciousness,
In a universe of matter and space,
Past and future, we won the human race.

It only took 13.7 billion years,
For these many rare events to occur.
Though just a few of the coin tosses were good,
The bad tosses went nowhere in a hurry!

Well, I've left a lot of good fortune out,
And perhaps you readers can help fill it in;
Know, too, that bad luck may come as well:
Global warming, nuclear war, or more asteroids.

The lure of myth is ever great to man;
But, beyond the apparent solidity
Of the word 'faith' is its meaning—of
Belief in the unseen supernatural!

Matter and motion manufactures all
Being and time in the arena of space,
We the complex composites from simple stuff,
The ultimate, so far, in the universe.

OUR WORLD

Two specs of dust met and stuck together.
This was the beginning of planet Earth.

MOONLIGHT SONATA

The music of the spring was in the breeze,
A prelude borne by airy musicians
Of the trees—the mating calls of the birds,
That opened for the cosmic symphony.

The Music of the Spheres played in the park
At night—flung down by our Father, the Sky,
Through the soft night to our Mother, the Earth,
Then to us, their audience and progeny.

The planets joined in a concert to the
Merrie Monthe of Maie, arrayed as follows:
There was Venusia, the Bringer of Peace,
Singing side by side with warring Marsius.

Flitting about was the wingéd Mercuria,
The speedy messenger who conducted
The orchestra, melting all of us who
Were touched by her wand of burning desire.

And mighty Zeus, was there, full to the brim
With the jollity of the fat man's belly.
By Jove, came Saturnus, so very gray
With age—lumbering into the party.

Thence sat Urania, the magician, and
The old sea captain, King Nep, the mystic,
But not Pluto; he was downsized, no more
One of the harmonics—an underworld!

Jupiter's music was round and robust,
While Saturn's boomed with sounds of grandeur
And the old venerable melodies;
But, Mercury soon picked up the pace.

Next flowed the serene love songs of Venus,
Followed inexorably by Martial marches.
Now was the time for Urania's magic—
She played musical jokes and surprises.

At last, their music came to mesh as one,
And our wanderers of the night floated
Away on the haunting mystical strains
Of King Nep's tune, into the May Flower moon.

Now we're touched, so touched by the starlight,
Afraid that we'll ne'er be the same again.
Can you sense the euphony of the spheres?
Can you fathom the theory of everything?

NEVER FAILING

Is it that we are
But a shimmering glitter
In the eye of eternity—
A small parentheses
Enclosing a dust mote
Of a rare and lucky event
Of little significance
On the edge of forever...

Or, that, as our luck
Has never ever failed—
Our joy and innocence
Will ever prevail—

Because we are
On probability's path
Of the ever possible,
As long ago worked out
In the realm of potential
That evolved this universe
To consciousness,
Which, as such,
Collapsed the wave function
Of all possible universes into ours?

THE PURSUIT OF MERCURIA

For some years I have pursued that lovely
Greco-Roman woman named Mercuria;
I've yearned till I could no longer reason.
Once, just her sight would have pleased me;

But now, at whatever cost and downfall,
I must taste of her fiery passion;
At whatever risk I plot her every move.
When the time is right, I'll be seeing her;

It will be just us, while the world's asleep.
The problem is that she's a fast woman
And is quite difficult to even sight,
Much less capture, entrance, embrace, and kiss.

And I can only have her for awhile;
Before dawn: if I linger with her long,
We'd soon be consumed by a rising fire;
After twilight, we'd be lost in darkness.

Yes, I have courted her many times,
But she's so elusive, fleeting, and small.
Once I waited for her just before nightfall;
All was perfect—'twas the best time of all.

There was the calm of a windless sunset,
Then the brief brooding of twilight's gloaming,
And the promise of a slow sultry night...
Clouds arrived—and so I missed her again!

She strayed not far from her fiery lover.
While I may have glimpsed her (I wasn't sure),
She slid toward her master's gravity,
Condemned to whirl about his light;

However, I was quite determined;
'Twas the thrill of the quest that kept me strong.

I planned to surprise her just before dawn...
I crept onto the frosty roof, near slipping,
There waiting. Damn! Clouds were boiling along
And blocking the view of her beauty rare.

Suddenly the clouds cleared, and she was mine—
Just over the eastern horizon was
The planet Mercury—dear Mercuria—
I stayed with her as long as possible,

Naked in the night, until, to blazes
She went when the sun arose; however,
Memories remain of those precious moments
And now she belongs to me forever.

Venus is too easy, Mars is always there,
Jupiter ever-present, Saturn bright,
The Earth under my feet, Pluto underworlded;

King Neptune, Queen Urania? Where are you?

WHAT FATE BECOMES OF LATE
THAT NEXT ARRIVES ON MY PLATE?

What has fate in store for me
And when comes the delivery?

What providence determines my destiny plucked?
Is it the stars, by chance, that rain down luck,
Or is it just serendipity and good fortune
Created by our own karma of kismet done?

What hands mold my future certainty pot,
Managing the outcome or the end of my lot?

Do the Fates 'Clotho, Lachesis, and Atropos' three
Predestine the preordination meant for me to be?
Am I doomed and bound? Is there any guarantee?
What do I care since all seems so free to be!

FROM TOE TO BEING
AND FINDING MEANING THEREIN

Why & How

Nonexistence can't be, nor could something
Make itself or always have been perfect,
For, before definition is the possible—
Timeless–formless—all options were open!

What, Where, Who, Then, and When

'What' matter stabilizes in 'where' space,
Begetting the appearances in motion as
'When' future moves through the 'now' to 'then' past—
This "spirit of life" granting our 'who' being.

The Forces

The strong force facilitates stability;
The weak force allows changeability;
Electric action, leading to magnetic motion,
Facilitates the movement of appearances.

The TOE to Being

The TOE has to explain origin, method,
And life, and, so, this does, the key being
That movement of appearances begets
Changes in time, showing in our life's realm.

Universal Answers

Since there's no rhyme nor reason for existence,
We're free to make our own meaning of it;
If we don't, then it's really meaning-less;
If we do—it becomes the ultimate!

Luck Happens

Asteroids swept away many species;
Two chromosomes fused, leaving chimps behind;
RNA remembers all survivors;
Good fortune smiled on Homo Sapiens.

The Balance Sheet

Life on Earth is death's borrowed debit;
We spend this life on good fortune's credit;
We're not God's puppets, but free of the strings;
Dispensing with angst, we're free in being.

We Are What We Are

Unintelligently designed, humans
Were a lucky accident of nature,
A haphazard Rube Goldberg 'invention',
With a nervous system ruled by ancient times.

The Lucky State of Us

As an accident of evolution,
We have the ultimate freedom of choice—
No "God's will"—we're beyond instinctive;
We're free to grow and evolve, through learning.

Difficulties Abound

Emotion often bypasses the intellect;
Many stand at the brink of insanity.
Only education can save the world—
We're at the turning point of history.

Wishful Thinking

Pride: Ego exaggerates self-importance
To claim that we're specially created,
Deserving a divine destiny.
Humility: we're electrochemical.

Unfortunately...

Those who can't or won't learn are doomed to stay
As their robot selves, living the sitcom life,
But, learning disperses the myths of old—
We make our own way or stagnate and die.

Meaning—or Not

Direction arrives or one goes nowhere;
Growth happens or one vanishes to null;
Creation comes or reaction destroys;
Planning makes a life or it collapses.

Coming Full Circle

Searching for the ultimate happenstance
Of how we began leads to exploration
Of within and without, a rewarding quest;
Upon return, we know the place for the first time.

THE POPE...

Just called me for some advice...

I told him to get rid of the celibacy rule;
Note that priests have a 17% rate of schizophrenia
While the general population has but 1%;
Ease up on sex not being natural, for it is;
Read Voltaire, and give up dogma,
For it states its truths all at once,
Only to have them fall one by one,
And that the charity stuff was great,
But get rid of everything else.

WE ARE MOST FREE WHEN
WE ARE ASYMPTOTICALLY CO-JOINED

The strong family unit, as the three quarks,
Is bonded by the power of its grouping,
But, loses identity if the home breaks—
Other pairs soon forming after divorcing.

Or comes the prison of solitude,
Chained to isolation with fortitude,
Floating, lost, without effects of affects,
Losing the identity conferred by others.

Within the proton, gentleness becomes strength,
For the members are free to explore at length,
Never smothering, but building unity,
The unit's direction adding to the one.

The strong force grows weaker near the quarks,
And so we may observe them someday,
Shining in their primordial glory—
The beginning of all things composite.

Identity is not lost in the co-joining—
True loves don't crowd the hearts of the others,
But, rather, look outward, in the same direction,
Close, joined, but don't block the others' section.

It is a seeming arithmetic violation,
That in summation we become greater;
We don't merge, having supported freedom,
Yet still share the same good vibrations.

Love matures when partners let it flow beyond—
Free to wend its way to places dear and fond.
Love's butterfly prospers when winds blow free;
Unconditional love never binds—it bonds.

THE SIMPLE TOE THAT EXPLAINS ALL

Before time was the timeless—all at once;
Before form was the formless—everything;
Before space was the spaceless—nowhere.

Existence had to be since something IS,
For nonexistence is an impossibility.

Quarks, photons, and electrons were of
Possibility, either a lot or a bit,
Along with dark matter, dark energy,
As what collects from gravity into stars.

The strong force stabilizes matter,
The weak force disperses matter;
Electromagnetic force allows movement.
Gravity arises from all of the above.

Stars generate the basic elements;
Supernovae generate all the rest,
And these atoms form molecules that lead to
Life's complexity—all from simplicity.

RNA tracks successes, failures meet death;
Luck, time, and accidents evolve the species.

Appearances in motion grant our being
Through their what-where and then-now-when.

The brain interprets reality and puts
A face on the waves of sound, light, color, touch,
And a sense on molecules' smell and taste.

Consciousness is the brain's perception of itself.

We operate through the where-to and where-from,
Using the what-then and the what-when of matter,
Leading to history, progress, wishes,
And remembrance, or being—

All made possible through the movement of
Appearances that creates past (then) and
Future (when) through the now, along with
Matter (what) in space's place (where).

So, then, the Theory of Everything after
Reality Was Generated is:

Being

=

Matter moving through space from possibility.

WAY WAY BACK

In 1909 in British Columbia, near the town of Field,
Walcott and wife were riding horses
Along a mountain trail
Beneath the Burgess Ridge
When his wife's horse slipped a stone,
Tipping and turning over a slab of shale.
He got down and looked;
There were fossil crustaceans unknown.

The next summer he climbed up the mountain's side,
Having traced the presumed route of the rock's slide,
And there he found a shale outcrop as long as a block
Imprinted with Earth's ancient and tiny livestock.

'Twas from the dawn
Of life's great and complex profusion
From so very long ago—
It was the famous Cambrian explosion.

MOON CHILDREN

The Earth would wobble like a dying top very soon,
Without the steadying influence of our lovely moon;
But, it's slipping from our grasp an inch & a half a year.
The end's not so near, but we'll need a way out of here.

THE HOLOGRAPHIC UNIVERSE

When a tree falls in the forest
And there's no one around to hear it,
Does it make a sound?

No, for there is no ear to turn
The sound waves into sound.

Nor is there a smell, for there is no nose
For the odorous molecules to attach to,
Nor has it any color, for there is
No retina to decode the light frequencies.

What does it look like, then?

It doesn't look like anything,
For there is no brain to put it all together
By detecting form, color, texture,
Size, taste, smell, or vision.

Since the entropy of a black hole is known
To depend on the surface area of
The event horizon and NOT on its volume,
Then our third dimension MIGHT BE a projection.

A projected illusion, as in a hologram,
May still be used as it were really there
Since we can make sense of it, so to speak,
But, in truth, the third dimension may not exist.

Thus, apparently separate particles,
Like created photon pairs,
Copy the other when one is changed,
Because, in truth, they are still
The same thing in the projector room.

If the universe is holographic,
Then the tree in the forest,
Whether seen or not,
Is, at heart, an interference pattern
Brought to life only when we tune it in.

This is the mystery of the realness
Of sleeping dreams revealed:
We tune in to the interference patterns,
Whether awake or asleep,
To bring alive the reality projected.

Everything connects to everything else
Through overlapping interference patterns,
And so nothing is so separate at all, as it seems,
But is one large all-encompassing whole.

Memory, too, seems to be holographic,
Residing everywhere in the brain,
Every piece associated with others related,
Instantly broadcasting all the connections.

Every part of a hologram contains the whole,
The whole universe contained within
A grain of sand, all eternity within a moment,
The universe rumbling when an electron vibrates.

We are part and parcel of everything—
We are the cosmos; we are life; we are love;
We are all that is; we are the creator
Of the dance as well as the dancer.

Whether the past is recorded and accessible
As part of the holographic whole is not known
Or whether the other two dimensions are
Projected, as well, but perhaps we shall see.

This then is the secret of the universe,
Knowing of that which underlies all reality:
Fundamental, absolute, indestructible,
Omnipresent, indeterminate, but all pervasive.

Why absolute and fundamental?

Because it is made of one piece—itself,
And therefore indestructible, and eternal, too,
And makes up all that there is, everywhere.

THE DNA OF THE UNIVERSE

The Infinite may radiate through a matrix,
Using Information or Energy to create
The Cosmic Background antenna which broadcasts
Interference patterns of virtual reality.
(ha-ha)

DIMENSIONS

-1. For making up dimensions
With no point,
For measuring the imaginary.
(The twilight zone)

0. The point of existence.

1. Going to great lengths.

2. Plane to see.

3. Eyes' cube.

4. Time or gravity.

5. Electricity

6. Magnetism

7. Heat

8. All possible universes

9. Heaven

10. Beyond Heaven
...

16. 7th Heaven

ENERGY MATTERS

Perhaps matter is more than just equivalent
To energy, if it were transformed,
And more than equal to it, more or less—
It may be that matter IS energy.

Perhaps energy is projected by
Information or is a part of our
Reality illusion, as well, but,
We can't stand on turtles all the way down!

Only four percent of the Universe
Consists of the matter we know and love,
The remainder being hidden from us,
So, perhaps, this is what's really in charge.

The basis of the Universe was forever here,
For nothing can make itself from nothing at all;
Such, a state of nothing could never be, for there IS
Something—something that consciousness interprets.

Mind and matter are made of the same stuff,
That "substance" made only out of itself.

Mind experiences the present moment;
Matter records the present from the mind;

That is, Present Mind, Past Matter, combine
The frames of Space and Time into the film
That lives and plays in us as Consciousness,
Mind taking Space and Matter doing Time.

Well, how can I save the Soul—Consciousness?
It may create potentials, quantum-like,
That give rise to the Reality of
The Mind and Body—so, use it wisely.

WHAT NO MAN HAD THOUGHT BEFORE

Alan Guth had never done anything much before,
But soon attended Dicke's Big Bang lecture tour,
And so he'd decided to study the birth of the universe;
Thus, just like that, he developed inflation theory first.

The "Big Bang" formed 98 percent of matter spent,
But, whence the rest of all the higher elements?
What flaming forge fired carbon, iron, and more?

Fred Hoyle was a nut, much unloved, a big bore;
Working with others who often avoided him,
Hoyle came up up with imploding stars, a whim
That that allowed supernovae to generate
The heavier elements at the rate of his steady state.

This process was known as nucleosynthesis,
Causing a 100 million degree heat and mist
That sprayed new elements into clouds of stardust
That could eventually coalesce into solar systems, us.

99.9% of this mass made our sun, the rest leftover dirt,
Ever colliding, two grains being the conception of Earth,
For, in every encounter there was always a winning lump
Of these endless and random, bumping, growing clumps.

(Fowler, not Hoyle, obtained the precious Nobel prize;
Hoyle had been overlooked, but to no one's surprise.)

THE ACCIDENTAL HAPPENING

What random, unsystematic event became,
So unmethodically, quite arbitrary, so lame,
Unplanned, undirected, so casual, uncausual made
As some indiscriminate, nonspecific one bade
Of haphazard stray that erratic chance gave?

CLUELESS

Let's not worry about a TOE that's sought,
But ever enjoy what it has wrought.
To me, happiness is a state of mind,
And so then we do much enjoy all that we find.

WE ARE

Existence is sure and real,
But its mechanics are no big deal,
For this life is the message dear,
Not the messenger.

FUTURE PERFECT

The trick of life
Is to foresee the past
By remembering the future.

EVOLVING

The mold of our jello was cast,
As the die was thrown as death;
Then we tamed the energy of life,
Remolding it beyond the strife.

We roll the dies, over and over, and try,
Then roll over and die—from snake eyes.

THE WILD AND VARIED ASSEMBLY

Absolute Consciousness made the party,
Attended by His poor relatives so hearty;
They bragged of their opinions so different,
Pro and con, canceling out to a Great Silence.

THE TRANSIT OF VENUS
ACROSS THE FACE OF THE SUN
AND
THE UNLUCKIEST MAN
ON THE FACE OF THE EARTH

Edmund Halley had suggested
That if you measured the passage of Venus
Over the sun from selected places on Earth,
You could work out the distance to the sun,
By using triangulation,
And then go on to use that calibration
To find the distances
To all the other bodies of the solar system.

These transits come in pairs eight years apart
And then there are none at all for a century dark;
There were none in Halley's lifetime, but in 1761,
Twenty years after Halley's death, the world was one.

Scientists set off for points all over the Earthly globe,
Hundreds of them, but most remained in problem mode;
Many were waylaid by war, shipwreck or sickness;
Then, too, there was much damage to the instruments.

Jean Chappe spent many months traveling to Siberia
By horse, sleigh, boat and coach, nursing his criteria
Over every bump.

At last he was near,
But swollen rivers blocked the way—
Locals blamed it on him looking at the heavens.

Guillaume Le Gentil set off from France
A year ahead of time,
But got delayed and was yet stuck at sea in brine,
Impossibly trying to take measurements
From a pitching ship.

He continued on to India,
Now having eight years to prepare
For the transit of 1769.

He erected a viewing station there,
Having everything ready on the fine day of June 4th;

But, just as Venus began its pass, a cloud slid forth
Right in front of the sun and stayed there and spent
Its time exactly for the the duration of the transit:
Three hours, fourteen minutes, and seven seconds.

Enroute to a port to head for home,
He contracted dysentery and was laid up for a year,
But then finally left the territory
On a ship that was later hit by a hurricane off
Of the African coast and nearly wrecked and lost,

But, he did make it home 12 years after setting off,
Only to find that relatives had long since sealed his fate
By declaring him dead and then plundering his estate.

The few measurements from 1761 were of no benefit;
But, luckily, in 1769, Cook had watched the transit
From a sunny hilltop in Tahiti, giving enough weight
Of information now for Joseph Lalande to calculate
The mean distance to the sun
At about 150 million kilometers.

NO PROBLEM

Another problem with problems is math class,
Wrestling with the equations so we can pass.

Ah, but's what's in an 'abstract' number? Naught
But the entire universe and the grade that's sought.

NEITHER THIS NOR THAT

If what wills the will is not done randomly,
Then determined it must then be,
But what of ties where there is no rather?
Well, one tie may wear as good as another.

WHAT SENSE TO MAKE?
WHICH PATH TO TAKE?

From what beastly heart springs our zest?
Of what searching eye became our sight?
What sound in the bushes let us hear?
What dark past haunts but helps us be?

Across what ink black river did we have to swim?
To what ends at length did we search for food?
In what deep entangled forest were we bred?

And hitherto,
Of what stars did we shine of their stead?
And in what nursery were those infants fed?

THE UNANSWERABLE QUIZ
OF ALL THAT SOUL IS

Whoever can tell
What's inside the shell
Of the soul and the brain,
Each not of the other's reign...
Which, pray tell, does what?

THE ESCAPE FROM SPACETIME

I had a night dream
In which I dreamt that I woke up.

Now I was once removed from spacetime,
But, just to be sure, went back to 'sleep'
To then dream within the dream.

So, in this double virtual reality,
Spacetime no longer mattered
And thus I had a whole 'nother life;
Everything under the sun was new!

ALL MUST BE, SINCE NOTHING CANNOT

Are there as many stars
As all the snowflakes that ever fell?

Of course there are, and more.

Does forever ever end, dying out?
Do unbreakable basics ever wear out?

Well, once every thousand years the Bird
Of Time flies over Mt. Everest, and downward;
On some of those occasions, a portion
Of a feather falls upon the mountain.

When the mountain has worn itself away,
The end of forever has thus arrived, that day.

ITSELF

All 'before' was open; everything was lawless,
Timeless, spaceless, and formless;
Nothing could not progress.

So, what needs no cause? Possibility;
For possibility needs but itself to be.

HUMPTY-DUMPTY

The egg falls and breaks, from heights too high,
But on the ground of being it rolls along.
The good eggs wobble, taking the dips and dales,
For within the shell grows a fertile embryo.

VAPOROUS

The spirit attaching to brain and sense
Would be immortal unperceptive non-sense.

1816: THE YEAR WITHOUT A SUMMER

In 1815, on the island of Sumbawa in Indonesia,
Tambora exploded, killing a hundred thousand people.

The Earth began to cool from the smoky ash and dust,
And sunsets became extraordinarily colorfully beauteous;

Lord Byron dreamed that the bright sun was dying;
The spring never arrived & the summer was very trying.

Crops all over failed to grow; Ireland was famished.
Earth's temperature had fallen by but 1.5 degrees F.

They called it the year of
Eighteen Hundred and Froze to Death.

RECEDING

The first cause could not be of Mind Aware,
For complex composite's parts must precede.

FUNDA-MENTAL

Fundamental parts compose the physical
And mental parts of the systems above
That are made of those basics below—
The complex composite movement of being
And mind—necessarily preceding them.

Awareness, thinking, doing, dreaming,
And seeing takes a 'village' of constituents,
Those always having to have come first.

We cannot just wave a wand to make it not so,
Even absolving God from it by exemption.

TIMELESS THOUGHTS

I threw my watch away,
So now I'm free each day,
As time cannot chain me—
Till the end of eternity.

I slept when I was tired,
And at my job got fired,
But my spirit went higher
Till my cooking caught on fire.
(So, I'll just keep my timer.)

THE LAST CHANCE SALOON (CASINO)

Entropy is always the winner in the end,
When there's no more money left to lend;
Meanwhile we stabilize, in nature's way,
Rearranging resources temporarily.

We gamble natural selection's bets,
Many leading on to a broker death;
But, we have a good supply of stable guys,
To put right back in when a player dies.

THE WORKINGS OF THE BRAIN

The mind is an apartment complex,
Some simpletons trying to take it over,
Until they are evicted from compartments
When the higher authority boots them out.

MINDING OR MATTERING?

If all is minding, then nothing matters,
All being illusion, a wisp of a dream...

But, if it matters, then it ever minds of that
Of which you are much concerned to be...

FINDING THE EDGE OF THE UNIVERSE!

At Princeton University, Robert Dicke and his team
Had really been building up much scientific steam
From pursuing George Gamow's good suggestion
Of a deep space Cosmic Background Radiation.

Gamow wrote another paper suggesting some ways
To use the Bell antenna, but no one read it in those days.

Unknowing of this paper and unbeknownst to Dicke,
Arno Penzias and Robert Wilson, but 30 miles not far,
Were diligently trying to get rid of this very CBR!

At Bell Lab,
Their large communications antenna deployed
Was plagued by some persistent background noise,
A steady steamy hiss, unfocused and unrelenting,
They ever attempting
To squash it away very painstakingly.

For a year they'd tried to eliminate this nuisance noise,
Through testing, rebuilding, and wiggling-dusting ploys,
Even placing duct tape over each & every seam & rivet.

They even wiped away a ton of bird shit from the dish,
Scrub brushing it and sweeping it clean; but, no fish.

Little did they know
They'd found the edge of the visible universe:
The very first photons were at hand—
The most ancient light,
Although time and distance
Had changed it into microwaves.

It was this interfering radiation
They wished to swish away.

If the Empire State Building was the universe we know,
They had reached within an inch of the sidewalk below.

In desperation, they called Princeton about the noise;
"We've been scooped!" Dicke sadly told all of his boys.

Penzias and Wilson received the 1978 Nobel Prize,
Even though they'd not been looking, CBR-wise,
And didn't even know what it was when they found it,

Nor had they ever described it in any scientific paper,
Not even knowing the significance of it,
But from the newspaper.

(Sadly, all that Dicke's team got
Was a bit of sympathy.)

Note: they didn't really call it "bird shit",
But a "white dielectric material".

See The Birth of the Universe At Home:

You, too, can detect the ancient CBR;
Just tune your TV to a blank channel;
About 1% of the dancing static is the CBR.
When there's nothing on, it's really everything!

VERY DETERMINED INTENTIONS

Of what are God's decisions made,
But of causes that beneath them gave,
And from what did these determinations bade,
And where were those foundations laid?

THE SHOW

The TOE is that we cannot KNOW, so far...
Having gone through many tries;
So, we're as free as the birds,
The bees, and butterflies!

COME AND GONE

Like the light from a star already spent,
Our 'get up and go' has long gone and went.
We all birthed, lived, and died right away—
There's nothing left but the slo-mo replay.

THE SNEAKER

Once there was a software engineer,
Who at home was seldom there.

One night he said he's going to have an affair;
"You can't fool me," his wife said,
"You're going back into work,
Aren't you dear!"

THREE CHEERS FOR STABILITY!

Energy manifests into existence
Of its movement that makes extant
The range of "particles' in time
That gain some amount of persistence.

It does nothing further to say
That His Energy was, too, this way,
It behind and before our own energy,
For energy is neither created nor destroyed.

IMAGINARY TRUTH

The "seems", "could-be's", and "maybe"
Migrate to "is", "must be", and "should".

Supply the proof and I will that refute,
Unless it is so that there is no proof,
But, if so, please supply the circumstantial;
However, if there is none, I'll take imagination.

THE OTHER SHOE DROPS

Determinism doesn't sit well, at first;
Its flavor does not quench the thirst,
For then it seems we but do as we must,
But, we'll see a way that in this we'll trust.

We wish that our thoughts reflect us today,
Our leanings, for it could be no other way.

To know, let us turn to the random say
To see whatever could make its day.

Shifting to this other, neglected foot,
What could make the random take root?

It would have no cause beneath to explain
It events, they becoming of the insane.

We could pretend, imitating air-heads,
Posting nonsense on purpose in the threads,
But that then we meant to do this way,
Noting history, too, so random holds not its sway.

There's less problem of a determined Nature
Than the same in our individual nature,
But, sense isn't made from random direction
That relies on naught beneath its conception.

Would we wish it to be any other way?
Doing any old thing of chance that may?

The random foot then walks but here and there,
Not getting anywhere, born from nowhere.

The unrooted tree lives magically, unfathomed.
Is not then randomness but a fun phantom?

The opposite of determined is undetermined,
The scarier ghost that's never-minded.

THE ACCOUNTING PRINCIPLE
BEHIND OUR UNACCOUNTABLE ACCOUNT

Nature's books did not balance,
Due to probability's chance.

On a 1 in 30 million shot
The roll of the dice was hot.

(Good that Professor Pat was not there, pals,
Closing that account on nature's thumbnail.)

There was a surplus
Which became us...

...To cash in
Before we cash it all in.

We would thank God for all us fools,
But now we see that probability rules.

SEA AND SKY

The Caribbean evening
Songs tucked in
The planetary paramours,

As Jupiter and Venus
Pulled the cover of night
Up and over their bed;

Then sunk the crescent,
Sideways into the sea,
But its two horns showing.

This rare sight of moon to see
Sent us into ecstasy,
While darkness brewed its tea.

THE ENDLESS PATHS OF THOUGHTS

Some speak for God or Evolution,
Showering those forms with
Imaginary Souls and Wills.

This vanity, willy-nilly thinking,
Assigns what is wished for,
But, thoughts make not facts.

Imagination wanders along
The infinite paths of this wonderland,
Not all roads being equiprobable...

...Coming up with contradictions
And complications.

To go where no one knows
Is, then, to go nowhere,
Through these forks of imagination.

These steps upon unsupported steps
Weave the threads of the
Web that ensnares the mind.

OF CAUSE THE FIRST

Whatever is eternal and is so well defined
Could never be as so, for it was never defined
In the first place, for that there never was
To define all that it forever did and does.

NO FIRST HAND

Falling for second hand
Words of the invisible Man
Is no more to understand
But of some accounts by man.

NOW HERE VS. NO-WHERE

What is here is to know;
No further else to go—
Neither there nor where.

Life calls; it must be lived;
We are happy to oblige
This existence before essence.

The imaginators preach;
We're out of their reach,
Where life can teach.

Our traits are all alright,
The gift of evolution's light,
The dark with the light.
We walk the shores of life.

NIL

Nothing from nothing won't do;
Nor something from nothing, too;
So, forever never began;

The tiny things yet arose,
In certain numbers chose.

NON-RANDOM

Isn't there determinism
As our catechism,
Or, if you will,
There's no free will?

We cannot other be,
So, farewell objectivity?

INTUITION

Of conscious reasoning
It is not always,
For the brain
Still works silently
On the problem assigned.

Subconsciously,
Not for one to 'know',
One's memories, learnings,
And associations interplay,
Sifting through the many
Scenarios of consequences.

Then, after this analysis,
The answer pops out
Into consciousness,

Wherein, sometimes,
We must truly be surprised,
Having not been privy
To what occurred beneath.

NON-FUN-DA-MENAL UN-ABSO-LUTE

What ever is apart, 'above', must, too, blow a fife,
But, as a first consciousness it can have no life,
For who's this highest Guy to which it All belongs
And from what other notes played His songs?

THUMBING

We hitchhike through the galaxy,
Our thumb protruding from our hand,
While, on the other hand,
We have a thumb and four fingers, too.

THE CHICKEN AND THE EGG

Which came first,
The mind or the matter?

To me it seems that
Matter needs not mind,
But that mind ever matters.

Yes, it still
Works both ways,
The material moving mind,
And mind moving material,

But, if the brain gives thought,
Then mind is material, too.

Now, if the mind be immaterial
Then what do they have in common
For this energy to exchange?

HOW MUCH ENERGY?

The whole of energy transforms,
And so it is never the less;
Some forms like protons appear stable
But who knows, they may change someday.

Of infinite energy, if there is,
It would be pretty crowded here;
But, if there were, some of it could die
And all would still of endless infinity.

NOWHERE

At that long gone doorway
Of where the origin once was,
I knocked onto just thin air,
For there was no one there.

THE ASTEROID THAT MAY DESTROY HUMANITY

The air beneath it
Couldn't get out of the way of the rock,
Rising in temperature ten times more than the sun is hot.

Everything & everyone crinkled and crackled in the heat;
A thousand cubic kilometers of earth blew from beneath.

This shock wave, radiating at about the speed of light,
Would sweep just about everything else out of sight.

From further away, one would see a blinding light,
Then the unimaginable grandeur of an apocalyptic sight:
A rolling wall of silent darkness as black as midnight.

It would reach to the heavens,
Filling the entire field of view,
Traveling far beyond the speed of sound
Toward me and you.

A bewildering veil of turmoil
Would [ful]fill our vision
During those few last minutes
Before we met oblivion.

WARP POWER

Our mental fabric quilted truths have long been sewn,
By evolution and whatever wove and woofed the known.
At first we admire but a few strands unknown,
Then blend the weave and weft to all its beauty shown.

DUST TO DUST

From naught or atoms we came
And to naught or atoms go our remains.

DOUBTFUL

It could very well be that
That which exists in the end
Exists also in the beginning,

That evolution was truly so,
[Actually] happening
Kind of deterministically,
Of stuff with a history,
Rather than a direct,
Movie-like presentation
Of now as an instant dream,
And, of course,
'Twas neither a Nothing
As formerly thought.

So, we converge,
Those ideas and mine,
Now describing all of [long] time.

But it could not
All have Been
Of an aware Mind,
For then we'd have to
Explain it away in
Much the same way—
Those endless regresses,
Then proceeding to infinity.

And so, perhaps,
Instead of a Mind
Foreseeing 10**1000000
Inexplicable contingencies,

And thus being
Of an even larger question,
There was a physical law,
One just beyond 3-D,
Quite naturally,
In which all possible universes
Played out, potentially,

Ours becoming the actual
Since it went the furthest,
Or even from consciousness real-izing.

THE POINT

Infinite points can be
On a line of here and there,
And infinite lines make a plane,
Then those infinite planes a cube.

Infinities of forms with no ends
Could never be completed in time,
So then they cannot be,
Of no maternity, or eternity.

ENERGY

Depending on the nature of energy,
It compresses into the 'solidity' of an atom,
At a certain point, upon whirlpooling;
But, ever trying to accumulate more energy,
Which becomes unneeded;
So, then, this excess is whirled off as photons.

If the nature of energy were different
In some other universe,
Then, too, the speed of light
Would be different there.

PRESCRIPTION

You can be made from what's in a drug store bin:
Carbon, hydrogen, oxygen, and nitrogen,
A dash of sulfur, a little calcium and a pinch
Of a few other very ordinary atomic elements.

IF ALL WAS A DREAM

The intricate eyeball lets not anything in;
It's only for show in this dream we're in,
As are one hundred billion brain cells,
There being nothing at all that they tell.

So says Mel, but this doesn't ring a bell,
Nor does the ear hear or a nose smell,
Nor any thoughts to think or know,
Nor post; it's all a dreamy magic show.

THE EFFULGENT

There is the light of the known all around
And so I find it of no use to draw down
The shades and turn off the switch to stub my TOE
In the dark upon the invisible undetectable.
— St. Austin Abdu'l-Baha

HERE

Oh, joy of life behold!
Of any cause uncaused!

Whatever became so old?
On existence I am sold!

WHO AM I?

Domain: eucarya
Kingdom: animalia
Phylum: chordata
Subphylum vertebrata
Class: mammalia
Order primates
Family: hominidae
Genus: homo
Species: sapiens

SCALING A THOUGHT

A trillion trillion atoms are
In a cubic inch of water—
Just some drops in the ocean.

Our solar system,
Is seven trillion miles across,
Including the multitude
Of Kuniper objects.

The extension to and of the Oort cloud
Is but a dot in this galaxy
Of a trillion stars,
That, too, are multitudes...

As among those stars
In billions of galaxies
In perhaps a multiverse of universes
During an eternity of time.

Lo, a spec of a vainglorious mammal,
Among 50 million species, or so,
Upon the earth, after 4.5 billions years,
Yet steps forth to say
That all there is was put everywhere
So that we may thrive for a moment,
Much else being for show,
It all being here with us in mind,
Granting our promise, in kind,
And illuminating us as 'special' humankind.

WHAT IS

Not being a wiz,
He guessed the answer on the quiz,
About what existence is, equal or not:
The answer was not his.

WHERE DID THE UNIVERSE COME FROM?

It can't be from nothing, for nothing makes nothing;

It can't be from God unless He is an ET,

For otherwise it begs—to have to then explain
All That with nothing beneath It;
Besides, He would still be part of everything;

Everything couldn't have been forever,
For how could a thing already made
Not ever have been made?

Ah, thus it all came from possibility—
And there's no for it, too,
Having become of possibility,
For it is already there.

Why? Because, obviously, nothing could not be!

THE LIVING FILM

Oh how I am moved by all that moves
In this glorious moving picture show
Of thoughts and images arriving on the stage.

Sometimes we take a front row seat
And other times we sit way back.

ICE FLOATS ON WATER

If water lacked the bizarre property
Of the title above,
Ice would sink and no longer
Hold in the water's warmth,
Which would then radiate the heat away,
Leaving it chillier,
And, of course, creating more ice,
But, it's not like that.

BILLIARDS

All of pool that they say is so true,
For my father bought me a table, too.
Being in the right spot at the right time
Is one's good fortune created on the dime.

After homework was all done, we'd make the run,
In billiards playing banks and carom fun.
We'd have to hug the girls to teach them how
Then even demonstrated 'kissing', wow!

After school, we too played some pool
in the smoky den, where gambling fools
Bet their lunch money time and time again,
Where, too, some girls mixed among the men.

I won an Army eight-ball contest,
Winning a carton of useless cigarettes,
So I gave them away, for the play's the prize,
Keeping me here and away from paradise.

I haven't played much in thirty years,
Nor have I drunk even that many beers,
And some long shots I still had to make,
But the short easy steps were best to take.

I turned to tennis, the theme much the same,
Though the bounce was but upon the ground:
Don't look up too soon from the knowing aim
And follow through—works too for other games.

Never played much golf, but did sometimes,
Driving was all right, but nothing really fine,
But upon the putting green, my eye was right,
Sinking putts like billiard balls, left and right.

ORIGINS

Evolution gave us a beauteous skin tone,
But, under that, we are ugly to the bone!

MEASURING THE SIZE OF THE EARTH

An English mathematician, Robert Norwood, of many,
Wished to know the circumference of the Earth, as any,
With his back against the Tower of London, he forked
Two devoted years marching 208 miles north to York,

Repeatedly stretching and measuring a piece of chain
As he went forth through all the heat, cold and rain,
And made many meticulous adjustments tolled
For the rise and fall of the land and the meandering road.

Then, in York, a year since he began in London,
He measured the precise angle of the sun.
Thus, using trigonometry to size a degree of the mark,
He came up with 110.72 kilometers per degree of arc.

Not thinking that these measurements could be true,
Since the slightest errors could throw them into the blue,
Jean Picard spent two years trundling and triangulating;
Using quadrants and pendulum clocks, he got 110.46.

But, was the Earth fatter at the north and south poles?
Now new measurements were needed to replace the old.

A hydrologist, Pierre Bouguer and and soldier,
Charles Marie de La Condamine, with many bolder,
Traveled to Peru to triangulate through the Andes,
To measure the meridian from Cuenca to Yoarouqui.

They needed but to go 200 hundred miles for one degree,
But everything began to go wrong, often spectacularly.
In Quito, they provoked the locals, getting stoned away;
Their doctor was murdered and the botanist went crazy.

Fevers and falls claimed even more; the senior member,
Pierre Godin, ran off with a pretty thirteen year old girl.
Then they had to halt their work for eight long months,
Having to sort out a problem in Lima with their permits.

La Condamine & Bouguer stopped speaking,
And all the officials had many suspicions, unbelieving
That the French would travel halfway the world around
To measure the world right here in their very own towns.

Why didn't they make the measurements in France?
Well, Edmund Halley, an exceptional figure, by chance
Got from Newton that our planet was slightly oblate;
But, Jacques Cassini had come up with the reverse fate.

Jacques erred, but the Academy sent the team in mind
To South America, to mountains with good sight lines;
But, the mountains of Peru were often lost in the clouds,
So they'd wait weeks to observe a bit, complaining loud.

The terrain was near impossible, defeating the mules;
The men plodded on, fording wild rivers, hacking jungles
And crossing uncharted stony deserts far from supplies,
Tackling the task for 9 long sun-blistered years of lies.

They then found out that another French team, cold,
Had taken measurements in Scandinavia that showed
That indeed a degree was longer near to the poles,
The Earth Forty-three kilometers wider equatorially
(Than from top to bottom around the poles.)

Still not talking, Bouguer & La Condamine just moaned,
Returning to the coast, even taking separate ships home.

WHENCE FORTH?

The thought had arrived, unexpectedly,
So how could s/he will what was already to be?

Whence did it bud, conditioned,
So silent, before it grew to fruition?

It bore all the hallmarks
Of one's memories and learnings,
In some new association.

RE-ATTACHED

Like two golden birds
Perched in the self-same tree,
"Lo," one said to his siamese twin,
"I am thee and you are me,
Forever joined in harmony."

The Ego and The True Self
Are such intimate friends
As to be inseparable
By any operation,
And so each may ever
Have joy unto the other.

The former partakes of
The bitter and the sweet fruits,
These experiences bursting
Alive against the palate,

A grand taste
Of the The Tree of Life,
While the latter looks on in
Glorious attachment and witness
Of life's wondrous experience.

LIVING

It is best to live everyday life
As it appears to be,
For the perceived state of being
Is not that beneath thee.

SEEING THE LIGHT

One need not look where there is no light,
For being alive is the very meaning of life—
Living life is the sparkle that gleams so bright;
Why parade in the dark when this here's so right?

THE KARMA OF THE BARKING DOGMA

Some Hindus, Buddhists, Christians and Jews
Wondered what stories they should choose;
Even thought they'd already so many chosen,
They just didn't want to keep notions so frozen;

So they met to merge the postulations into one,
Thinking that this might be a whole lot of fun.

"In our hypothesis, there is just the Only One."
"Well, our conception is a multitude of many Some."
"Well, we'll partway meet: there's only the Holy One"
"Nah, the odds of that are over three million to one!"

"Buddha of us was one, so of Gods there are none;
A human above all that now's never seen by the sun!"
"Humph! Holy Jesus of our one God was His son!
He lit mankind's darkness with light of the Sun!"

"No, Jewish Jesus was not of any nature Divine,
But was just a mere man much ahead of his time.
This you all should know, being there at the time.
Look at our history singing those biblical rhymes."

"All is not real, so what is this great big fuss?
Retreat back to where it's all at to slow the rush."
"Oh God's universe and creatures are so real
And that is why we're making this very big deal."

"In the afterlife, we in Hell or Heaven reside."
"Not so fast, for in between these realms we lie,
And if you in this testing life don't do so well,
You'll have so many subhuman tales to tell."

Reason arrived: "Possibility reigned way then back
'Before'; there's nothing even holy about all that.
'Tis all made up, those many fabrications made,
So just let it all be, for this is what existence bade.

THE LAST DODO WORKED IN A MUSEUM

The famously flightless bird, the good old dodo,
Had a dimwitted but ever trusting natural motto.
During millions of years of isolation from us,
It had evolved on the island of Mauritius.

It was not at all ready for human behavior low,
Even waddling over to note the fall of its fellow.

In 1755, seventy years after the last dodo' s word,
A museum director at the Ashmolean in Oxford,
Nothing that its dodo specimen had become "tired",
Being unpleasantly musty, so he threw it into a fire.

We are now not even sure
What a living dodo was like,
But for some oil paintings.
We will not again see their likes.

'TWAS POSSIBLE

Once upon a timeless time
There was no reason nor rhyme,
Nor any form of form the less,
Nor any laws for the lawless.

So what could underlie Ye?
Yes, 'twas all just Possibility.

DEATH IS A WAY OF LIFE

Of all extinctions, the Permian was the largest.
245 million years ago, for 95% of animals perished,
Suddenly disappearing from the fossil recording.
Life had almost come to a total obliterationing.

MASTODONS AND EXTINCTIONS

In the late 1700's, Cuvier could take heaps of bones
And whip them into shapely forms not in the stones.
After describing and naming
The fossil elephant the mastodon,
He put forward for the first time
A theory on extinction.

He said that from time to time
There were global catastrophes
In which some groups of creatures "became history".
This raised uncomfortable implications at the time,
For why would God create and destroy
Without reason or rhyme?

This suggested an unaccountable
Casualness by someone unseeing
And greatly troubled the belief in
The Great Chain of Being,
Which held that the world was carefully ordered for us—
And that every living thing thus had a place and purpose.

Meanwhile, William Smith noted a correlation in fossils
In rocks to find the relative rock ages that were possible.
At every change in rock strata, certain fossils vanished,
While in others they carried on into subsequent levels.

Now it was seen that God
Had wiped out creatures extinct
Not only occasionally but repeatedly—
Which made us think
Him not only careless
But having an outright hostile distinction.
There had been more than
The Biblical Noachian deluge extinction.

HOW CAN WE FIND PARTICLES?

What it really takes to find particles today is money!

NEVER MIND

The first cause could not be of Mind Aware,
For a complex composite's parts must precede.
So, think no more of things before some things,
But of 'before' existence and physical laws.

THE RICH AIR OF THE BIG TIME

Back when oxygen was more
Plentiful than now we know,
Dragonflies grew as big as ravens,
For the heady "O's"
Had encouraged outsized growth.
Horsetails and tree ferns grew to fifty feet,
Club mosses to a hundred and a third.

NO FIRST REASON

Every-thing, Every order happens for a reason.
Yes, for the most part, for most seasons,
But not for the bottommost cause the first,
For there was nothing before it to reason it forth.

TIMOTION

I feel so much lighter now since I learned that
The Earth's expanding beneath my feet, and
Pushing me up, raising me onward, for—
Just knowing this gives me a lift!

The road now rises up to meet my step,
And I no longer fall down 'drunk' to it.
I've added dimension to my old self
By sending "time's motions" forces through it.

THE QUALIA OF LIGHT WITHIN THE DARK HEAD

Photons arrive as some electromagnetic waves,
As do the vibrations of undulating air waves,
Yet no sound nor light does out there tread,
But is transformed to such within the head.

THE EVIDENCE OF ATOMS' EXISTENCE

Who was it that provided
The first incontrovertible substantiation of Atoms?

Twas Einstein in his paper on Brownian motion.

SOUL FOOD FOR THOUGHT

We bless the 'needed' soul with the holy kiss
Of life, being this of which to replace us with;
For what did natural selection ever do, in vain,
Spending so extravagantly on the higher brain?

Well, I declare, I see hearts that pump the blood,
And all of the chemistry born of that great flood,
As well as cells all about for everything human,
But wherefrom of the same is thought's acumen?

Because I make of this a mystery, as those before,
I'll suppose the answer here, that and nothing more,
And say that an invisible soul infuses us, running us,
So that we can know all of that not here before us.

THE SPIN ON THE SPIN

We revolve, rotate, turn, and go round the sun;
We whirl, gyrate, and circle our most loving one.
To TOE we wheel, twist and turn, twirl and sing,
Ever swinging, swiveling, pirouetting, and pivoting.

BACTERIA:
THE BACK DOOR TO OUR STOMACH'S CAFETERIA
AND THE INVINCIBLE RULERS OF THE EARTH

For two billion years in the Archaean world, bacteria
Were the only forms of life. Algae, or Cyanobacteria,
Learned to absorb water molecules, dining on hydrogen,
But releasing oxygen as waste; photosynthesis began.

The world began to slowly fill with "poisonous" oxygen,
But not right away, as it first combined with iron then,
Producing iron oxide that sank, that on the bottom lay,
In primitive seas, the world literally rusting away.

After 2 billion years, the atmosphere had some oxygen;
A new kind of cell arose. Some oxygen-using organisms
With organelles produced an energy much more efficient.

This was the endosymbiotic event of a mitochondrion
Which made complex life possible, by a liberation
Of energy from food, feeding on nutrients we take in.

We need them but they don't need us, for without them
We couldn't even live for two minutes.
They don't even speak the same genetic language
As the cells in which they live.

These eukaryotes are old and unknown visitors
Within our homes who've stayed on for a billion years.

In another billion years they learned to form together
Into complex multicellular beings, yet, still this world
Of the small was to ever live on and rule the world.

At dinner, Louis Pasteur used a magnifying glass for
Searching for microbes in his food, until invited no more.

There are 100 quadrillion bacteria within us & upon us,
Ever grazing on our flesh and digesting our food bus.
The Earth is not our planet, but theirs; they let us live.
They even purify our water and keep the soil productive.

A single bacterial cell can generate 280,000 more a day.
They can also share information, taking a piece away
Of genetic code from any other any time. They swim
In a single gene pool—an invincible superorganism.

They live in caustic lakes, in Antarctica, in boiling mud,
And even thrive seven miles down in the Pacific Ocean;
In sulfuric acid, too, and in a 166-year-old bottle of beer,
And can even gorge themselves on plutonium nuclear.

Bacteria were yet alive in a sealed camera lens stowed
On the moon for two years, but they seemed a bit slowed.
Some were even found two thousand feet below the Earth
Dining on what's in rocks, like iron, sulfur, and dirt.

Some frozen ones were even revived from the 3 million
Year-old permafrost of Siberia, and even one older than
The continents, was resuscitated, a 250 million-year-old
Bacterium that had been trapped in a salt deposit hold,
Two thousand feet underground in New Mexico, maybe.

POTENTIALLY

Heading towards the future of infinity,
Dissolves the past eternity
That could have no maternity,
And thus reforms it retroactively;
So fear not the haunts of the present.

PALINDROMES

Too hot to hoot!
Never odd or even?
Too bad – I hid a boot.

Madam, in Eden I'm Adam.
God saw I was dog.

— Nitsua Austin

THE CALDRON THAT ALMOST
BREWED HUMANITY AWAY

At Toba, in northern Sumatra, a supervolcano
Erupted only seventy-four thousands years ago.
Six years of volcanic winter followed this eruption,
Bringing pre-humans to the very edge of extinction.

There were but a few thousand of them left around,
Since very little light could reach the dusty ground.
It took twenty thousand years for them to recompose;
From this handful of hardy souls we humans arose.

In 1960, Bob Christiansen looked around everywhere
At Yellowstone National Park for its volcanic caldera,
But found it nowhere. By some coincidence, NASA
Had photos from a recently tested high altitude camera.

Astounded, Bob learned
Why he'd failed to spot the caldera;
It was virtually the entire park,
2.2 million acres of area!

Yellowstone must have blown up with a violent misery
Far beyond anything known throughout our history.

The crater was forty miles across. The cataclysm was
Even beyond the scale of what the imagination does;
It had thousands of times more monstrous molten fire
Than Mount St. Helens. Krakatau was but a firecracker.

Yellowstone's eruptions average
One really massive blow
Every 600,000 years,
The last one being
630,000 years ago;
It is long overdue;
Better take out
Some no-fault insurance.

ANSWERS

Science discovers the truth everywhere;
Philosophers just sit around in chairs;
Religion just makes for bigger questions;
Evolution explains how we got somewheres.

INFORMATION

Learning provides more deciding power,
Enlarging the spectrum for choice to flower.
I surf the internet for that which matters,
The world's bouyance there the former latter.

THE GOLDEN STREAM

In 1865, Hennig Brand thought that gold
Could be distilled from human urine, old,
Perhaps noting a similarity in color,
So, he kept fifty buckets in his cellar.

By some method, he converted urine
Into a noxious paste of some kind,
Then into some translucent waxy substance,
But so far there was no gold, and none hence.

However, after a time the substance began to glow,
And when exposed to the air burst as an inferno.
The substance soon became known as phosphorus,
But was too costly to make its business prosperous...

For an ounce of the flaming stuff sold
For way more than the price of gold!

UNBOUNDED

I was certain that I was bound to be,
Ever predestined since maternity.

NO MORE OF THINGS
COMING BEFORE THINGS

How could that the First be?

My name is Possibility.

Yep, just little old me
Was all, you see,
Around for eternity
With no maternity.

What came before me
Was only but possibility,
That being plain to see.

So, all of you, hear ye:
Couldn't all nothing be,
Nor substance for free
Already ready made so wee.

Weren't no laws: that's me,
Before the physicality.

All was lawless; no He,
I made substance be
Because I have no fee,
Nor could naught be.

Out plopped a little quark
Yet alone in the dark;
Then another spark,
Every one a lark.

They made protons tight
Throwing off some light
As photons going right,
For once having might.

This success was probability.

THE ROOT OF ALL EVIL

Other than direct hurts to persons,
Which are covered by civil laws
Is what some groups think of 'good' arbitrary?
And harmless, until it is imposed on people?

We see many good and bad things directly,
Person to person, via the actual,
And such are the good civil laws
And good human values taught.

The problem becomes when we 'see'
From no direction but the imagined,
Via the unreal.

These 'good' things, merely pronounced,
Also define their 'bad' counterparts.

One then 'forgets' their null source,
Leaping into their complete adoption,
Becoming more and more with them one;
Thus, the ideas must be protected.

Anger arises toward the contrary,
As emotion stains the brain.

Then, evil is done
In the name of a concept of 'good'.

All these 'good' things
Eventually
Come to a bloody end.

TOGETHER

It's tough for men and women to exist in isolation,
For, the nature of one makes necessary the other.
A good way too find yourself is to lose it in another;
However, it becomes rather a shared identity
That does not destroy the identity of the other.

IMAGINARY FRIEND

Humans wanted a larger role and model,
So we painted our feelings on God's will.

Infinitely magnified, He crashed our skies;
With friends like that we need not enemies.

I guess we created God as Frankenstein,
As our will had to do, natural selection
Evolving us to look for intent first and
Worry about the constitution of it never.

THE DIFFERENCE

Science is of that which can be known, by definition.
Science shows little variations in what is known.

Science loves to find increasingly refined
Or even different answers than were expected.

Science is for the inquiring mind;
Science is of the visible and measurable.

Religion is of that which can't be known, by definition.
Religion is also compounded by contradictory variants.
Religion(s) already have every answer,
More like "that's it; case closed";
Religion is for the limited mind.

Religion is of the invisible and never measurable.
How, then, is the twain to ever meet?

BIRTH OF THE SOUL

It's easy to pronounce and declare the wish
That a soul does that and a soul does this,
But the soul has nothing which to do it with;
So we give it a mind, heart, depth and width.

REPLY

As you say, Vincent,
Of conscious reasoning
It is not always,
For the brain
Still works Silently
On the problem assigned.

Subconsciously,
Not for one to 'know',
One's memories, learnings,
And associations interplay,

Sifting through all the
Scenarios of consequences.

Then, after this analysis,
The answer pops out
Into consciousness,
Wherein, sometimes,
We must truly be surprised,
Having not been privy
To what occurred beneath.

WHERE IT'S NOT AT

Going back down the scale, though the atom
And on into the quark or string or whatever—
To the simplest point—is certainly not
Where we would expect to find
The most ultimate complex composite
Of all time and size to the infinite power.

THE UNWELCOME GUEST

We are all relatives, descended from stars
And trees, as in coming down from.
Some of our relations are many times removed,
But they just keep on coming back!

OTHER WORLDS

Beyond our planets
Whose names and order are indicated by
"Martha visits every Monday
And just stays until noon, period"
(The 'a' of 'and' is for the asteroid belt
And the 'period'—Pluto—is gone)
Are the exoplanets
Of other solar system,
Discovered only since the 1990's.

We really need
A more specific spacecraft
To get better looks,
But some "hot Jupiters"
Have been found
Too close to their stars,
And so they must have migrated there.

...Maybe some kind
Of Planetary assembly line...

DE TOUR

I took a tour
Of some exoplanets,
Ranging from Heaven to Hell...

On GLIESE 876D,

With an orbit
Tighter than Mercury's,
I waited an earth-year
For the sun to come up,
Since it rotates so slowly.

I noted its moon,
Its atmosphere shredded
By solar winds.

Sunrise released
A fiery Hell,
So I stepped ever back
Into the twilight dawn
Of a blood red sky.

On Tres-4,
There was an airy feeling,

It having the density of balsa wood;
And could literally float on water.

No one can explain why it is so large.

It is perhaps but a toy
In Someone's swimming pool.

TW Hya b
Is a hulking baby,
Only 10 million years old,
Compared to the 4.5 billions years of Earth,
But, it's ten times the mass of Jupiter.

Thank God for GPS.

55 Cancri
Is ever bountiful,
It's binary-star system
Hosting five known planets.

It lies within
The habitable zone,
Conducive to life,
So, perhaps,
I'll retire there.

IMMATERIAL

It is true that thoughts
Can come out of the blue,
The brain ever churning
On its learning.

There is a trust, too,
That this is still of you.

I see Jimbo's energy
As transforming,
Spinning, condensing
Into material solidity...

This energy being
As real as in E=MCC...

Some people, though,
Use 'immaterial' differently,
Such as in soul or floating 'consciousness',
In which case
There's nothing in common
With the 'material',
Unlike our real E=MCC,

And so it was these, so different,
Could not talk to each other,
To exchange energy,
Each being not of the other.

However, using Jimbo's energy
Now as the new 'immaterial',
It, As Jimbo says,
Had to come before protons,
As they are spun of it,
And, of course, before
Stars, atoms, molecules, cells, and brains;

Then they do whatever brains do,
Such as making thoughts
Of a subconscious analysis,

Giving over the result
300 ms later to the mind
To surface into consciousness
To witness as experience.

The question still remains, though,
Of what 'loose' unattached energy can do,
If anything, but cause static and noise
Or have no effect until it amounts
To something material.

SATURN'S CRYSTAL HEXAGON

On his cell phone, Graham texted a telegram,
Which is really what texting is like, damn,
With it's new codes and abbreviations
(Which is too long of a word for its definition),
On out toward the sexagon on Saturn.

Graham needed more space
From his wife-said place,
So to the ice palace hex,
He went to have good sex.

Back later came a mama-gram from the mams.
Those lovely ladies there said for Graham the man
To come back, bringing quickly some quarks
For their quirks...

Came another tell-a-Graham from those mams,
Saying, Master, we six slaves awaiting
Have just delivered those many sequels
Of that last and happy Saturnalia's ringing bells.

Graham went off to attend to these family matters
On Saturn concerning his sexagon,
Although he still claims that his sex is a-gone.

WHICH CAME FIRST,
THE MIND OR THE MATTER?

To me it seems that
Matter needs not mind,
But mind ever matters.

Yes, it still works both ways,
The material moving mind,
And mind moving material;

But, if the brain gives thought,
Then mind is material, too.

Now, if the mind be immaterial
Then what's in common
For energy to exchange?

THE WORD[S],
THE LEANINGS AND GLEANINGS

Where in the Woe is Purgatory's bane?
Purgatory's on Venus, where sulfurs rain.
Where in the Heck is that deep Hell of pain?
Hell's found in the sun's heart, oh, hot burning pain!
Where in the name of Heaven is Paradisea?
Of Heaven's site, no one has any idea—
Really now, where's Heaven one and the same?
It's the world's best kept secret: Earth is its name!

Yes, that's said, but truly, where is the stead
I must tell of them that they're only read...
Of those places spent after we are dead?
It's written of words that language bred.
'Twas hope-ward(s) that invented all that was said?
'Twas these that were signed for anything Divine ["said"].

MID-POINT

Twilight dawn or dusk are
The still points of the noise,
The day-night neither here nor there,
But in equipoise.

LIFE IS BUT A DREAM?

Of all that seems there is nothing there;
It's all a dream from we know not where.
'Tis so much nonsense made of thin air,
For the Dreamer Guy is having a nightmare!

It's not real, all this we feel to be,
But is just a phantasm of reality.
This would be plain for all to see
If there was anyone here to see.

That we can see makes us have to be.
This reality is as stable as it can be,
An amazing state for a 'virtuality',
So what then would the difference be?

If nothing is then there is no quiz,
No right or wrong from a testing Wiz,
So I'll just remain the same as His,
Living out this earthly dream, as is.

NO MIDDLE OF THE ROAD CAN THERE BE

We can't just sit on a fence but in theory,
As in practical life we must answer the query.
All answers are not of equiprobable search;
So, caught in the lurch, we don't go to church.

THE GHOST NOT IN THE MACHINE

All aspects of one's amazing mental life,
Every thought pattern, every emotion rife,
Every memory and association made to be
Can be tied to brain physiological activity.

Cognitive science has shown that feats
That were formerly thought to be the feet
Spurred by the mental stuff of a soul lone
Can be duplicated by machines alone,

And that motives and goals can be
Understood in the terms entirely
Of feedback and cybernetic mechanisms,
And that thinking can be understood
As a kind of computational mechanism,

A kind of fuzzy analog to parallel computation.
So intelligence, which formerly seemed miraculous—
Something that mere matter could not possibly
Accomplish or explain—can now be
Understood as a kind of computation process.

THE STUFFING THAT
THESE DREAMS ARE MADE OF

The virtual reality of this dream that we feel
Is so much acting the same as if it were real
That any difference would not be a big deal;
Actually, there is no difference, so, it's real.

WHITHER FLOWING FREE,
ALL FROM NOT KNOWING

Of hitherto, I know not, but am whither going,
Willy-nilly, whence all there is to knowing...
Hence thither I went on hither flowing to find
That I was truly free to be in body and mind.

THE MILKMAN

What shaped the bottle that spilled the milk?
What intent or accident made it fall or tilt?
Of what prior substance was it built?

Well, we must drink it to the hilt,
So who cares how it was spilt!

We then sought the COW as the TOE,
But then discovered the bull.
...
At least we live fulfilled,
The elixir's already spilled;
So, it's up to us to enjoy the mess,
Ever becoming of our humanness.

FREE MILK

The guy bought the cow,
Ever keeping it here and now.
But other varieties of milk came out,
So he shopped around, that lout!

THE SELF-LESS RUSE

Mystic and Mysters ravel the TOE threads
With those answers that can never be read;
So many fights—the posters are spent;
Thence their descent, hence toward oblivion went...

TRILLIONS OF ASSORTED ATOMS MAKE YOU 'YOU', THEN THEY ALL GO OFF TO DO OTHER THINGS

Your atoms don't even care about you here;
They don't even know that "you" are there.
Your particular arrangement exists but once,
And you have only about 650,000 hours hence.

THE ONE TRUTH YET HOLDS FIRM

The most astounding knowledge we can know
Has come out of our search for the TOE:
It is for now that we cannot fully know.

We float many little balloons against this wall,
But it moves not, nor does it budge at all.

We wish so much that we could know
How everything came about that is so.
Our wishes are but small breezes that blow
Toward the edge of forever, but nowhere go.

It doesn't compute "God", this need to know,
Versus the reality of our present row,
Yet, the wishes remain and grow
Into presumptions heaped upon the "no's".

And so some I know, of those here on the quest,
Deeply and humbly see that our species and the rest
Have no special place here in this entangled forest.
As for me, I cherish every moment of existence blessed.

EVERLASTING GLADNESS

Happiness is a way of life that celebrates
A living aliveness—that then opens gates
To further adventure, friendship, and delights,
To joy, success, triumph, and greater heights.

THE ELECTROCHEMICAL ORGANISM'S 'SELF'

There is no unspeaking ONE of any proof,
For, meditation's FEELING that's of no self,
Nor bound is... neurological quietus.
While healthful, induced oneness is not TRUTH.

VACATION PLANETS

Uranus is quite pleasant compared to Pluto.

If you've ever had a dog, you know what I mean;
However, the under-worlded canine has been
Banished from the house of Astro—
To reign as the under-world in the Underworld,
For it's much better to reign in Hell
Than to be an unwelcome guest in the heavens.

Once, I was down on Venus,
And the sulfurous emanations
Were so repulsive that any gases from Uranus
Would have been to me as a breath of fresh air.

The gas giant planets' breadth and width is staggering,
And their mooning around is getting out of hand.

That leaves Mars as the only other good place—
Since Klingons have now appeared
On the rings around Uranus.

THE HUMAN MAMMAL'S MAGIC

The brain patterns reasons from the truths learned,
Texturing the conscious interface dynamically
For awareness to color our expressions of will
To accomplish what plans apply, by discernment.

VITALITY AND EXUBERANCE

With pep, zing, zip, oomph, vim and vigor,
He bounced along with spirit and fire.
Enthused by life's spirit energy of his zest,
He knew that this life was one of his best.

THE ENTRANCING DANCING

They were all dancing within love's treasure vault
Within the framework of a broadening thought,
The lights pulsing and the waves reverberating,
Where the good times had become everlasting.

Tribal primal field currents were raging
From speakers of the energy matrix pounding;
They whirled and twirled as loving gestalts
Of sentient consciousness knowing no halt.

There were rhythms of constant contraction
And expansions of bosom-energy projections
Converted to scalar waves of blinking attraction
As fission and fusion beckoned the connections...

...Ever forming in this Omni-sound emporium
Where tone waves vibrated in waves of creation.

THAT EVER UNDERNEATH

Oh Great Designer, burning nigh,
In the spaces upon high,
What energetic hand or eye
Could frame thy fearful symmetry?

In what distant deeps or skies
Burnt the fire of thine eyes?
On what wings dare You aspire?
What the hand dare seize Your fire?

WHEN YOU ARE YOU NO MORE

Who is concerned about an nonexistent brew,
A nonconscious state when there is no 'you'?

No one, for none spent the billions of years
Before they were born in a state of tears
Of anxiety and apprehension of not being here.

THE CURIOUS ENCOUNTER WITH MADAME

De Broglie declared that all motion
Of particles must be associated
With the propagation of a wave.
Einstein then wrote that De Broglie
"Had lifted the corner of the great veil."

Einstein later had an opportunity
To lift another veil—that of Marie Curie,
When they vacationed together,
Quite reactively, in the Swiss alps.

Did they or didn't they exchange energy?

Einstein was a ladies' man, though married,
Busy having an affair with his cousin, Elsa,
And Marie was a married man's lady (Paul's).

Einstein wrote his wife that Marie was a grouch,
But, was this just a misdirection meant to allay?
They inhaled the alpine air, talking science,
Strolling far and trying to name the peaks.

THE ONLY COMMANDMENT

Ever back through the Ages went I,
Dating rocks and old fossils, by and by,
And found this tablet stone, the Covenant
Of the one and only engraven Commandment:

NATURAL HISTORY

The Commandments of Evolution are unmistakably
Engraved in stone for everyone to see.
There are no "if's", "and's", or "but's" in these pages,
For we can even date these rocks of ages.

"WHAT IT IS LIKE TO BE ANOTHER CREATURE"

What would it be like to be another kind of creature?
Do I-thoughts of self-consciousness emerge
From their integration of experiences?

Who knows about the 'I', but, in a sense,
We already know about being other creatures,
For we already have been during our development:

The growth of a human parallels and recapitulates,
At a vastly accelerated rate, the evolution of life.
We start out as a single-celled organism,
Much like an amoeba or a bacterium.

Then we progress through the phase of a blastula,
A simple, undifferentiated multicellular stage,
To become an embryo barely distinguishable
From those of many other animals, even
Including those of reptiles and amphibians.

For these first few weeks after conception,
We are truly a lower form of life ourselves,
Bathing, as long ago, in the warm amniotic sea.

So, how did it feel to you? Can you recall?
No, for you were not around at the time;
There was no developed conscious sense of self.

But, now we do feel like someone,
Having inner depth,
Our activity and actuality
Being one and the same.

REPLAY

DNA is another proof of evolution.
It remembers all the successes,
Then plays them back during gestation.

LIFE EXPERIENCE

Before more surmising,
What are we to make of our experiences,
Whence and wherever they spring,
For here we are, receiving them into
Awareness/consciousness.

Which way to go?
Angst? Or acceptance?
Are we free of strings? Or puppets be?

Do we live on Good fortunes's credit,
Our life but a borrowed debit repaid at death?

We are forced to choose,
The right choice, if any, unknown,
For the mind/brain's answers are arbitrary.

Only conscious awareness is an indubitable truth
And it only receives thoughts as experienced,
Still not knowing if they be true or false.

We come into this universe,
Willy-nilly, not knowing,
Our lives given to us to live,
Willy-nilly flowing.

Yet, we are here in some way, no doubt,
Be it real or a very good imitation
That cannot be told apart from the real.

So, maybe, why not, just be, as is?

(—)

This represents birth through death,
Our dash between the parentheses of eternity.

GRAVITY EXPLAINED

The Strength of Gravity, the Feeble Apparent

Gravity is a *universal* force—for any body:
The force felt by a body is mass proportional;
Yet, the acceleration that's felt is the inverse!
This coincidence removes all mass dependence.

(Einstein transcended this amazing "coincidental" race
By bodies going straight through curved space.)

Gravity might be derived from the fundamentals,
The byproduct of a small residual after cancelations
Of opposite electric or color charges, and more.
Why then is gravity *universal*, for its sources are not?

Perhaps the appearance of feebleness is deceptive
Since protons and neutrons are but lightweights.
But why are they so light? Their mass is a compromise
Between a disturbance energy and its cancellation.

The quarks' color charge
Disturbs gluons around them,
Small at first,
But larger growing farther from the quark.

These disturbances cost energy,
But, how to cancel them?

With an anti-quark
Or two complementarily colored quarks.

But, the qualifying quarks
Can't sit atop the originals,
For quarks have no definite position,
Just a wave function,
And they can't be localized
To a small spread of position,
For this requires a larger energy;
So, forget nullification.

The compromise is that
Some residual energy amounts
From the not-completely-canceled
Gluon field disturbances
And from the not-completely-canceled
Quark positionings;

Thus, the proton mass from m=E/cc,
With this tricky element
Of how the gluon disturbance field
Grows with distance.

The residual strong energy
From color charge also binds
The protons and neutrons
In the atomic nucleus;

The electromagnetic electron/nuclei
Charge residuals
Bind atoms into molecules,
And molecules into materials.

Asymptotic freedom
Is a subtle feedback effect
From virtual particles
Antiscreening the color charge.

This antiscreening builds up gradually,
Especially at first,
Then proceeds more quickly,
Building upon each building.

Whereas, screening happens
For electrically charged particles,
Being such as a positive charge
Attracts a negative virtual cloud.

Thus, at first,
Since it's so slow to build,
The pressure to localize the nullifying quarks
Is quite mild as well;

Thus, there's no need
To very strictly localize and
So the energies are small;
So then is the proton mass.

This is the lightness of being.

(Ideas herein were gathered from readings,
esp. Frank Wilczek)

STAR VOYAGE

Once I was young,
(I said to the stars)
For once I was you.

I love your flames
Beckoning me
Toward the fires of home,

But I have moved on,
As all things must,
To atoms and molecules,

And on into cells,
For now I am them—
A glorious complexity.

I travel the long road
Thankful to all those
Who came before.

One day I may go off
Through interstellar space,
Bringing forth these tales to tell.

SEGNO (SIGN) # 5

"There's the aether", replied the other, apace,
"The 5th state, one that pervades all of space,
Yet there are no signposts of it up ahead,
Or anywhere, since it's in every stead.

"We regard it as the stuff of which Gods are made,
That lively spirit of elixir that their nature bade,
For, just as all mortal creatures inhale the air,
So do immortal and divine natures inhale the aether."

"The intimation is the mark of their manifestation,
A demonstration and a token of the evidencention—
The aetheric and heavenly sign of things to come,
Both the portent of the miracle and its omen.

"It is of the warning and the notice let,
Presaging both the promise and the threat.
Of this sign the aether follows, the gesture beckons;
'Tis the signal, the wave and gesticulation reckoned.

"We can read the writing in the sky, the marquee,
Daubed with symbols marking the cipher free,
With characters, figures, and hieroglyphs of time,
The ideogram of the rune, the emblem of the Divine."

SOL

Of stars, those lights of dark eternity,
Is one that now shines bright for you and me;
Photons race the sky across, shedding light,
Enlivening, illuminating humanity.

THE FRAME THAT WORKS

There is no time for space, no space for time,
They being of a separate reason and rhyme.
Time is a then, now, or a when of what is placed
And replaced in that place that is spaced.

IT HAD TO BE

'Something' had to be,
Called Potential/Possibility,
Since Nothing could not be;
Nor could 'something' be forever
Designed without any design.

However, this Potential is not a thing,
For a thing could never have been around forever,
Since that thing, with no beginning,
Would have already been here for an infinity of time,
For, things are chained to time, with no escape,
And, so, an infinity cannot have been completed,
For there would always have been
More time to extend upon it,
Not to mention that the real things
Would have worn out by now;

So, it was Potential that 'existed'
Outside of time, space, physical laws, and things,
It being the unconstrained spaceless,
Timeless, formless, and lawless superposition
Of this possibility of all things that are makable,
Perhaps operating all at once, or progressing 'rapidly',
Until the consciousness of organisms in one of those
Evolved possible states brought real things forth,
Consciousness being the ground
Against which events can be known,
Although not itself the events witnessed;
For, these are of our lives and being.

Consciousness, even now, with the mind-brain,
Still utilizes potential possibility,
By collapsing scenarios of consequences
Of actions into a surfacing thought or two—

And so it is that Potential was all there ever was,
Not quite as in 'forever', since it's outside of time,
But as still All, and in the sense that it HAD TO BE,
Since something IS HERE.

Now, in this reality that we exist in,
Movement of appearances grants time,
The appearances existing as/in space,
Being of greater and lesser densities,

Of which come wishes,
From thoughts of future space,
And remembrance from past space,
With history being of past things,
And progression toward future things.

ON THE ROAD

She loves road trips; the autumn colors called,
So we were off on the ups and downs,
She with taped ankle and myself with wrist,
The warriors running away from home.

The scene was of the turning leaves falling,
Unspoken poems reciting the paths flown,
Only now the scene painted with the words,
As music played poems sung to melodies.

Country roads, quaint inns, dilapidated barns;
What's this? A dance hall lighting the dark path?
We dance the song of evening bells rung
In a twilight zone in nowhere's middle.

The music played, past, but not yet past,
For it was in recent memory recalled.
Newly savored sensations continued on—
That which could be presently known.

Mind anticipated the coming tones,
The transitional 'middle' blending it
With those sounds not totally gone.

In this past-present-future resides
The delight that none could produce alone—
The smoothly rolling 'now'.

THE SOLIDARITY OF THE CONCORDANCE

The blend of the coalition grows upon itself,
Striving for the dynamic-balance—of light
And dark, Yin and Yang, and wrong and right.

Reality's not found in separate actions,
But in related events blended in twilight.

The concept of Classicism accentuates
Order and clarity of thought, simplicity,
Restraint, balance, dignity, and
A mistrust of emotion and excess;

However, since it relies on imitation and
The acceptance of objective standards,
It may lack spontaneity, and degenerate
Into excessive traditionalism and empty formalism.

Romanticism embraces an exaltation
Of the feelings, an individualism,
With new modes of imagination,
Of freedom of form, spontaneity,
Self-expression, and subjectivity.

It began, at least in art, music, and literature,
As a revolt against 18th century doctrines
Of restraint, forms and rules, decorum,
Stagnation, and blind tradition.

However, romanticism and classicism
Are now taken as more general terms.

Some exemplars of their contrast are:
Passion as opposed to reason;
The whole against the details;
The Yin facing the Yang;
The right vs. the left side of the brain
"Don't confuse me with emotion"
Or "don't confuse me with facts";
The sails confronting the rudder of the soul.

This epitome may become a battlefield,
Or it may grace a smooth sailing ship.

How easy they are not transformed,
These apparently opposing forces
That may wage war upon the other,
But, how tremendous they can be
In a bond of confederacy.

Pure reason, ruling all alone,
Is a force confining and stale;
While passion, unattended,
Is a flame that burns
To its own end.

Poetry is an ideal of the unison:
The right side of the brain
Provides the inspiration;
The left side devises the rhyme.

An utter, absolute classicist
Or romanticist is an extremist!
S/he honors one worthy guest
In the house above the other,
And so loses the love and faith of both.

Witness the average classicist at IBM,
One who knows little of the humanities,
One who ever works through lunch,
Never having the time to hear of life,
Making every decision by the book
But none from the heart.

Or the total romanticist:
One who can't even hold a job,
Even taking drugs, and losing all control.

The writing of this page—this analysis—
Is rather a classicist undertaking.
But, I do not live by the unbending way
And therefore my songbird
Is never imprisoned within.

Perhaps, it chooses to be here, classically,
Or perhaps it will, at any time of day,
Burst forth and enjoy a total feeling.

Nor does a long wild night of lovemaking
Mean that you've gone bonkers.

Life is full of spikes of valleys and mountains—
It is only when one can't merge the two
Or at least make jumps between
That one may need some reflection.

How can there be any sort of resolution
Of a dichotomy in which one side
Expresses itself so logically and
The other in emotions and images?

Well, if either your sails or rudder be broken,
You will soon be dead in the water...

Therefore, the discord and rivalry
Of one's elements must become
Rhythm and all sweet melody!

It's not the same for everyone,
But the knowledge of
The 'contrast' itself is the first step...

Therefore, let your blended soul exalt
Your reason to the height of passion,
That it may sing and fly about,
Letting it direct your passion with reason,
That your passion may live and survive
Through its daily death and resurrection, but
In effect, ever arising from its own ashes.

Now, no one can ever achieve
The ultimate and perfect balance
Between classicism and romanticism,
But for the rare times when in the 'zone',
And indeed, this balancing attempt
Itself smacks of classicism!

And so we all have leanings—
And that's what I mean when I say
My tilt is toward romanticism.

Emotion, slightly favored, rules,
But every so often I do check in
With logic and analytical reason.

Thereby, I enjoy the world, mainly,
Because, like many of you,
I am much impressed by it wonders...

Without perception's deeper depiction,
One finds little that excites—
Not noticing much, ever in a hurry,
And seldom having the time...

Two other poor relatives
Of classicism and romanticism
Are substance and surface glory.

The romanticist in me likes the veneer
Of the shiny red car or motorcycle,
But the classicist in me would like
To know that the vehicle operates well,
And even be able to take it much apart,
For that is the very substance.

When I maintain my car or cycle well,
Shine it up, and then speed off
Into the country sunshine
With the wind on my face,
Then I have the best of both worlds!

Now, I really don't know all the answers—
I just like to tug at the hem of the garment
In which life's mysterious dualities are clothed.

As ever as in all good marriages,
"The oak tree and the cypress
Grow not in each other's shadow".

People involved in the arts may
Like to listen to music while they work
In order to deactivate the left side of the brain
By giving it something innocuous to focus on.

Personally, I often dream up many ideas
While listening to music that moves me deeply,
For then the imaginative power
Of the brain's right hemisphere
Is free and inspired to soar unbounded.

Yes, I do lean toward romanticism...
Perhaps it is my nature nurtured,
Or perhaps I feel a need to counteract
The overabundance of classicism in the world,
Or perhaps because in romanticism there is grandeur,

While in classicism there is but cold logic
And endless analytical thought.

But, even with these leanings,
The good romanticists never forgets
That it is classicism that pays the bills
That authorizes the indulgences.

I have some hope, that,
In any totally classical person,
No matter how stern or dull s/he be,
That one day, somehow, somewhere,
There will come some small measure,
But, then, an ever-luring triumph of jubilation.

Yes, the desire to be orderly and factual
Is a part of the human species,
But there are other yearnings in every person,
The desire to be imaginative and unrestrained in
Expressing personal emotions,
Warmly and freely flowing,
And to take in art, music, literature,
As well as escalate the way one lives a life
From an illuminating flame fed from the self,
A source of lucid experience that

Can usher wisdom and fervency,
As the means to the rounded truth.

Then luckily, these may be some of its aspects:
Sentiment, celebration of nature, interest in the past,
A new emphasis on feeling and the senses,
Even actually enjoying melancholy and sadness.

Thence comes love of freedom, mysteries,
Even fascinating figures and heroes,
The allegorical, a delight in whimsy,
The improbable, and the 'impossible',
Of legend, folklore, and mythology,
An awe before the immensity of what is—
The Earth as a friend and
The sky as a warm blanket,
And certainly the uniqueness of the self.

The curious blend never lets one down,
Ever keeping one centered, but ranging.

So, extroversion entertains, at large,
While love's introversion is great, one-on-one.

Intuition and sensing
Can sustain each the other
In a magnificent fusion.

Thinking and feeling combined
Are of an unbeatable synergy,
Of a being coalesced and intermixed.

Sensing the general direction but
Not exactly knowing one's next move
Is of a spontaneous higher 'order'.

There looms the classical planning of
A magnificently grand adventure,
Whether triumphant or of glorious failure
Always of the superb and the sublime.

Merge these ingredients, until smooth,
This loving mix, mingling and combining,
Soon melding into the 'zone', well integrated,
Stirred, whisked, and folded in and out, the commingling
That leads to the harmony of amalgamation's union,
The marriage and the synthesis, the very admixture
Of the concoction of life's ever-during brew.

The parts all sum to the whole flow, so,
Life must be more like a mosaic done
Than some focused laser tunnel of sun.

Since few lengthy pleasures are lent to us,
We build a stained-glass window of small ones.

Oh, thou soul, dare to live near the edge;
Brave the walk of the line, balancing fun
There between adventure and misfortune—
For the greatest blunder in life is to
Repeatedly fear that you might make one.

Hail! Lord Byron's Golden Mean extends:
Let us have wine, lovers, song, and laughter—
Water, chastity, prayer the day after.
Such we'll alternate the rest of our days—
So, on the average, we'll make Hereafter!

Wholeness arrives by mixing the suspension:
Classicists drone toward dull perfection,
Romanticists drown in feeling's affection;
Worse, others alternate between extremes—
It's not this nor that, but a joined direction.

Harmony then rolls along, round and round:
Each holding within it the seed of the other—
Yin reaches climax, then retreats in Yang's favor,
A cyclic movement of rotational symmetry:
Rounded life is the blend of Yin/Yang together.

The perfect balance may still call upon us:
Edges dissolve when opposites are balanced—
Time and dimensional space are transcended.

Everything joins yet remains as itself,
For what "is not" is as great as what "is".

I SEE

The raw scene in the eye becomes seen
By the visual system,
That lovely painted face placed
For the 'I' to know.
or
The scene seen is eye to 'I'.

OH, GOD!

No, I'm not really shocked that God is a notion
Of the mind's conception, an abstraction
Formed into a hypothesis that becomes a belief
And a conviction, born of opinion's impression
By a higher mammal species having imagination.

HOLEY CHEESE

Holy cripes, we look to the holy skies: jeese!
The universe is but made of swiss cheese,
As our the moon must have been, no doubt,
Since it's crusty and hard from leaving it out.

BRAIN WIRING

Well, we knew that a little learning was dangerous,
And now thinking, too, for minds can be read through,
Well, my thoughts are sane, so it's OK to read a few,
And the rest of them I'll put into poems profoundest.

THE INTRINSIC INSTINCTIVE:
THINKING ABOUT THOUGHTS THEMSELVES

What is this conviction, in many,
That innate sense of impression fond
Of those spirits invisible and beyond?

Who or what put it there,
Those notions of the thin air?

To investigate, one must put aside
The very judgment that descends
From the conclusion deep-rooted,
For the inherent blocks its own analysis.

Whence it came forth so prevalent,
This indwelling urge to believe?

The plot ever thickens and twists
And turns upon itself, bare—
Natural Selection put it there!

One can have many feelings that surface
From the heredity of long ago.
Some are not so good, obviously,
Some are even forbidden thoughts.

Life's still emotionally primitive—
Some 'negative' feedback mechanisms in
The central nervous system, some useless,
Still send thousands-of-years-old messages.

And so the feelings are banished,
But subtle is the difference
Of these and those inklings
Closer to the boundary of distinction.

We don't fall for thoughts of violence,
Usually, although it is possible for some
To hear these directions as gospel;

But, we may fall for some 'innocuous' views,
Slipping over the threshold, indiscriminate,
Saying, "Well, I felt it, so thus it must be so."

Do we control our thoughts or do our thoughts
Control us? Could we, silly as might seem,
Just be falling, hook and line, for the thoughts?
Think deep—thoughts may tell you the answer!

We may fall for our thoughts, hook, line, and sinker:
Conditioned responses, reflexes, or
Overwhelming emotions, some spurious,
Or ancient, planted by evolution, or unbalanced.

Emotions are slow to react to logic,
Like molasses or slow forming crystals,
Or not at all, like rocks, blocking us.
Unless and until they change, progress halts.

Reason and emotion are hard to coordinate,
Each having a separate pathway to the mind;
That perhaps is all there is to tell about the
Miseries and follies of human history.

From its safe subjective place that's free of fear,
The higher self, our Conscious Awareness, can witness

The strange thoughts and emotions that surface
On the mind, sent there by the subconscious brain.

First-level thoughts are beliefs and desires,
But second-level thoughts are beliefs
And desires about the beliefs and desires,
Becoming able spectators of the scene beneath.

Higher Awareness, which can but witness,
Is a safe haven from which to observe
The drama of our lives playing in our minds,
Granting us a sobering distance from it.

This detachment allows this
"Thinking about a thought"
Without the thought itself
Trying to steal the show.

INTO THE BIN WE GO

The universe is but a massive Bingo game,
Rumbling the tumbling of our lives' gain.
We all sit upon the church basement floor,
Ever asking: what been go TOE? B-2? B-1? B-4?

THE BAG OF CHEMICALS THAT IS US

Do molecules of atoms, being 'chemicals',
Seem not to be for much, as those in the lab
Looking like so much slime and mold, making
One think that brains they could not constitute?

In what jelly blobs of meat do thoughts fly?
What the willful forge that flares us higher?
Upon what anvil are feelings pounded out?
As now we think of this, our brain neurons fire.

LUSH AIRS

"What is the Earth with these pastimes so fine?"
"It is the gift of the Universe's fine sweet valentine."
"What loveliness brings such soft breezes that caress?"
"The winds are the pressured airs mixing up the rest."

THE NATURE OF PANTHEISM

Panthea, the greatest God there never was...
How to explain? She does what nature does.
As a rose is still a rose by any other name,
Then so is a universe a cosmos the same...

THE SUPERTOE

Nothing can become of Nothing,
Therefore something was eternal.

There was no point at which something
That was around forever could have been defined,
For there was nothing prior to it but itself;
Thus, there was nothing to give it specifics.

Therefore it had no laws, no form, no mind
No space, no time, and no real definition;
It was causeless.

'Twas 'nothing' but a fluctuation
That gave rise to all thereafter.

The conserved energies of the universe
Cancel out, summing to zero,
But for the quantum uncertainty.

Our so-called material is but of the secondary,
And, thus, non-elemental emissions
Of opposite particle pairs.

All this, such as its other representations
Of positive mass-energy and negative gravity energy,
All sum and cancel to a big zero, literally,
Being but the arbitrary and fleeting phantoms
Of some temporary particulars of specifics
Sprayed from the uncaused, eternal,
And fundamental and necessarily
Indefinite ground-state beneath
That could have no real intent or direction to it;

For there could be no "before",
Nor of Nothing, nor of causes
Beneath causes of infinite regress.

Complete freedom is the glorious result,
Within our form, of course, with no strings attached,
Much better than a constraining "purpose".

WHAT IS MAN (AND WO-MAN)
BUT SAPIENS SUPREME

Oh Man! What a piece of work, the mind;
What noble deeds done and undone in kind.
What Rube Goldberg inventions heaped upon—
In the layers of brains the mind is made upon.

What is this sapiens mammal animal,
But of some slime and of brutish law.

Let us 'neglect' this state of affairing
On the grounds that it is unappealing.

So, then...

We are spun of the Eternal Golden Braid,
Those windings of Truth, Love, and Beauty made
From the Goodness of Purity Immortal—
The Theory of Everything's singular portal.

What is Man but the special chosen species
For which all the plants grow and the waters reach,
For which the Earth turns 'round, and orbits
A nuclear furnace spreading Love's energy,
Enabling us to thrive above any and all creation.

What is Man but the only bloom for which all
The 13.7 billions years of evolution and love
Have occurred in a predetermined random yeast
To form and flower such a vainglorious beast.

It's ever on forever's edge that we meet our destiny,
That in our temporary parentheses of Eternity
We would flourish for just a moment, bidden
As the blossoms of Perfection's Flower Garden.

A hundred trillion stars and countless shores
Were built to light our universal nights explored;
Forty million other lower species, too, the All-Might
Placed about our world, merely for our delight.

Our name is Writ Large on the Heaven's marquee,
In the supernovae stardust showered from Thee.
From Nothing not You came, but, of a naught
Our own universe was made and ever wrought.

A starring role we play in this reality show,
Every atom spinning fine just for us to know,
Our ancestors rising/falling for us to stand upon,
Oh man! They lived and died for our lone promise!

Every shaft of light shines with us in mind;
Thus, it beams forth our beginning and our end—
In and of God's hidden and Heavenly Shrine.
Oh life! We cherish being, that of yours and mine.

We do so much deserve reward beyond this role—
And so it is that one's immortal spirit-soul,
That angelic vapour that drives a living being,
Shall go forth to glory on behind the scene.

We are not merely some mammally organic luck,
But purposely evolved on this planet, near a star,
In that intended long and winding mindless 'birth'
Of slowly drifting time, dust, and selection by death
That ever sifted the best from the rest: Sapiens!

(Now why is the soul so 'true' and so far with it faith goes?
It is only because one so much wishes it to be what knows.)

Earth could not answer; nor the Seas that mourn
In flowing Purple, of their Lord Forlorn;
Nor rolling Heaven, with all his Signs reveal'd
And hidden by the sleeve of Night and Morn.

Ah, Love! could thou and I with Fate conspire
To grasp this sorry Scheme of Things entire,
Would not we shatter it to bits--and then
Re-mould it nearer to the Heart's Desire!

— Omar Khayyam

THE KNOWING

"Into this Universe, and why, not knowing,
Nor whence, like water willy-nilly flowing:
And out of it, as wind along the waste,"
Omar "knew not whither, willy-nilly blowing..."

Now I'm knowing, that out of this muddle,
Indeed, it's the chaos that frees me to be,
For, it's all of disorder in disarray,
An ultimate disorganized confusion,
Whence all sprung, banged, and exploded,
With no hint or trace of order, law or plan;

'Twas mayhem, bedlam, and pandemonium,
Wreaking havoc upon the turmoil of a tumult,
Heaping high upon, a commotion of disruption,
In the utter fullness of the uproaring upheaval...

...The maelstrom to end all messes and shambles,
The lawless free-for-all of total energetic anarchy,
Entropy crowned as King of the great hullabaloo,
That cosmic hoopla from which all hell broke loose.

Never there was to punish one for not even knowing
Why you are here in this world so much growing,
That became here all so willy-nilly going.
So, as life's rose, outspread your fragrance blowing!

Whither flowing free,whether knowing, or not,
Hitherto, I know not whence, but am whither going,
Willy-nilly, hence that's all there is to knowing...
Hence thither forth I go on hither flowing to find
That I was ever more free to be in body and mind.

It is of Ovid's "rude and indigested mass:
The lifeless lump, unfashion'd, and unfram'd,
Of jarring seeds; and justly Chaos nam'd.

"No sun was lighted up, the world to view;
No moon did yet her blunted horns renew:

Nor yet was Earth suspended in the sky,
Nor pois'd, did on her own foundations lye:

"Nor seas about the shores their arms had thrown;
But earth, and air, and water, were in one.
Thus air was void of light, and earth unstable,
And water's dark abyss unnavigable."

So it is that we the living might hereby agree,
To live a being that is much more intense,
To leap toward higher orders of actuality,
To revel in the glories of this conscious life,
To attain each minute a more euphoric joy...

And to bring this radiance forth to all,
The increased intensity of free experience,
And to build on it, etc.,
Ever growing;
Forever, amen!

THE SECRET

The basis of the Universe was forever here,
For nothing could make itself from Nothing at all;
Such, a state of Nothing could never be, for there IS
Something—a reality that our being interprets.

This then is the secret of the universe,
Knowing of that which underlies all reality:
Fundamental, absolute, indestructible,
Everpresent, indeterminate, and pervasive.

Reality now pulsates, in a real structured sequence.
A field that's present throughout space immense,
Out of which all particles can condense—
Occurring where the field's extremely intense.

Atoms are those bundles of inertia,
The knots in the field and fabric of space;
Yet matter defines the structure of space...
So the Yin is in the Yang, and vice-versa!

THE SUPER TOE IS CAUSELESS,
THUS, THAT IS THE SUPER TOE!

Our train of thought has driven us to the answer,
Of all that borne from near 'nothing' onto eternity,
Of the origin of the original disorder,
The lone dawn of our trackless radix,
Via the rails and tunnels that ever ran out:

There cannot be ever more and more
Causes beneath even more extended causes;
Therefore, intuitive or not, the causeless is,
Being such as what we observe it in the quantum.

Thus, cause is only of our higher realm,
As downward thence to its root emergence—
'Possibility' needed no mother but itself,
An egg burst open, born without a chicken.

The causeless bottom is the potential
Of possibility that is/was ever there.

Since it's 'defined' as an undefined chaos,
There's no problem of no initial definition had,
Since it can't have one and so it needs not any.

Things themselves become and go of 'virtual' potential,
Some things remaining as the rather-enduring real.
The potential is as near to simple as it gets,
Second only to the nonexistent Nothing, of course.

So, then, the potential is of no mind or 'seeing',
For that thought system can never be constituted,
As there are no more fundamentals upon more;
For, the Potential is already the ultimate basis.

Simple things ever combine, and further up,
And/or go must through phase changes,
Leading to more complex composites/forms.

Nothing, not existing at all,
And not even being able to,

But, perhaps threatening to,
Is the simplest state of all,
So, it must ever jiggle about,
Manifesting as loose 'change'.

You might say, then, that, that is exactly why
There had to be the potential for things;
Otherwise... Total Nothing, forever.

We have now reached the unexpected TOE,
One that even satisfies the ongoing trend,
For, looking down, we've always observed
The ever descending simplicity of Nature.

Now, as such, we can't really expect to find
An Ultimate Complexity sitting
Around there at the simplest point.

We didn't find Mind there;
Thus, we are ever free to be.

This causeless bottom 'fate'...
Was/is, too, a 'magical' state,
For anything could become of it.

SUPER-PARTNERS

The asymmetrical beast is a wonder to behold,
As within that wretched form is a heart of gold,
Beauty can be skin deep in one ugly to the bone,
So let there be uncle without anti being home.

ALL IN ALL

The nervous system connects to the brain.
When you feel a pain it is known in the brain,
Relating the sensation to whence it came,
So, it is all but an extension of the brain.

THE NATURAL HAPPENING

Why these little subatomic things, ask the wise,
In such amounts and of their special size?
Well, I agree that this shows they had to be made
As we see is from the causeless quantum's shade;

It's been shown by Aspect there's nothing hidden
Underneath this necessarily indefinite disorder
Whence random particles become, unbidden,
For causes beneath causes would have no border.

So, from this causeless bottom where bucks stop,
Hails the ultimate freedom to live and be a lot.
This scheme, undone, hints that the Ultimate Yore
If it could ever be, would need cause all the more.

Yet, at this very point, which is not an answer,
But a call to think no more and surrender,
Religion introduces Complexity Infinite
For the downwardly simpler bottom unit.

So, there's no answer given, but only
A larger mystery of the One and Only
That is an infinitely larger question there,
Rendering the entire 'answer' beyond repair.

While both science and religion claim the causeless,
They are as opposite as could be, none the less,
For one finds no specifics there, none at all,
While the other imagines a Perfect Ordered All.

If all the above opposed were not bad enough,
There is entirely insufficient evidence for God;
Zero, in fact, in the face of the opposite there,
For One who is supposed to be everywhere.

Beyond that of even the absence of evidence
For the interceding Ruler, the obvious nonpresence,
Which leads to the sure evidence of absence,
Is: that a first cause can have no reason to it.

Plus, that we observe the random emanations.

Humanists [non theists] push science forward
God naturally flunking out, with no pushes backward,
While creationists, with nothing to push forward,
Ever attempt to push science backward.

This, then, is the end of faith's season,
Being the celebration of logical reason.

THE UNDESIGNED

UnTruths are but great amusements and fun,
As we are the undoing of the Perfect One:

Adam and Even Sapiens goofed in no time,
As of Intelligent Design there was no sign;
Noah's progeny screwed up right and left,
Since they were of a Master's hand bereft;

Of the Ten Commandments they weren't impressed,
As no such thing came down from the crest;
Two thousand years of folly now from redemption
Were no picnic. 'Design' was from evolution!

Evolution, driven by natural selection,
Is a design without a designer.

THE BRAIN IN A VAT

Here I reside in a vat of molasses, king of the tub,
Getting tanked and enjoying the cistern,
For life is a barrel of fun that's unlimited
By my cask; Here I drum in my basin
As a precious vessel, living way beyond
My receptacle, the container not containing me,
But serving as my holder and my reservoir.

SCIENCE AND THE OLD/NEW
INTELLECTUAL CULTURE

From the 'snare and quark' to All-ology

The essence of science
Is conveyed by its Latin etymology:
Sceintia, meaning knowledge.

Science itself is then
The body of knowledge obtained
By using suitable practices for its fields.

Science has spread into many areas,
Even psychology and the social sciences,
And has become essential,
For science is the most accurate way
Of obtaining knowledge
About anything and everything.

Scholars who spurn science
End up with inaccurate results,
Such as Marx, Freud, etc.,
And they and the religious scholars
Preach *blah, blah, blah*—
As it's all up in the air, empirically ungrounded.

The traditional intellectual is being replaced,
For science-oriented investigation now renders visible
The deeper meanings and states beneath our being,
Redefining who and what we are.

The arts and science are now combining
Into an enlightened 'third culture'
Of a new intellectual landscape—
A Reality Club of the new humanists.

There are revolutionary developments everywhere.
The wonderful whole-like approaches,
Such as those of the long-gone giants
Of Leonardo, Newton, Michelangelo,
Darwin, and Einstein encompassing all.

The previous incomprehensible humanism
With an ignorance of science is fading fast away.

The previous culture that dismissed science
Is soon to become a fossil of the past.

These self-referential disciplines go nowhere,
Being most often concerned with
The exegesis of earlier thinkers,
In which one reflects on
And recycles the ideas of others,
With no real expectation
Of any systematic progress.

They just get further away from reality.

Science poses questions to elicit answers.
And the more science you do,
The more there is to do.
Reality is the final check and balance.
There are no fixed, unalterable positions.

Life plays an ever greater role
In the future of the universe.
Science is involved in all the humanities now.
Subject matter is discussed, not intellectual style.

Scientists talk about the universe,
Unlike many old style humanities academicians—
Who only talk about each other.

Those disdaining science
Are doomed to be left behind.

Certainly, human nature is fixed,
But its behavior isn't,
For it is sensitive to the environment,
Being endlessly variable and diverse.

Change the environment for the good
And behavior will then improve.
There is no real need to fiddle with genes.

The fixed rules of human nature
Can give rise to an inexhaustible range of outcomes.

To know what changes to the environment
Would be appropriate and effective,
You have to know the Darwinian rules.

We only need to understand human nature,
Not to change it.

So, something radically new is in the air:
New ways of understanding physical systems,
New focuses that lead to our questioning
Of many of our foundations.

A realistic biology of the mind,
Advances in physics,
Information technology, genetics,
Neurobiology, Engineering,
The chemistry of materials;
All are questions of great importance
With respect to what it means to be human.

(By some coincidence, after I made this,
'Scientific American' restated the theme
And has therefore listed the ten recent
Great scientific contributors to humanity.)

THE RUNAWAY MIND

Sometimes the mind so much wants to do it function
To know all, that it speculates its way to 'truth',
Not realizing that its mere pronouncements
Just float in the thin air as unsupported beliefs.

GOD: THE NON ROLE MODEL

'Tis lucky for us that God doesn't exist,
For in breaking the rules he'd ever persist.
Even His own commandments wouldn't be sacred
Since he'd murder His own forms created.

Well, this would be goof, big time—a mistake,
So then a joyous rainbow He might make,
To show He'd no more make a worldly lake,
But, He could still destroy us all by earthquake!

He'd slay by flame and flood excruciate;
He'd entrap; he'd blame us for His mistake;
He'd hold grudges for our ancestors' sins;
He'd throw tantrums and fits—his name, God's Sake!

Other loves would not allowed by this jealous One,
For He'd want to be the only one to enjoy the fun.
For His low esteem our adoration would be required,
This request being much like singing to the choir.

Would He have to sleep or rest on the 7th day,
After working 24-6 on making the universal hay?
Or would He have boundless energy reserves,
Such that He could do it all through an instant blurb?

Would God's last name be known as 'Dammit',
With 'Harold' His name on Earth's planet,
And would be 'Art' named—when up in Heaven?
Would we swearest in vain these names never taken?

We'd have to be really lazy on the Sabbath day,
Not even lifting up a finger or even wave a bug away,
Keeping holy and wholly the laundry on Sunday,
Even avoiding football, as the Pope might say.

Cripes, He'd have been in the right place at the right time,
Not ever having been made, not even costing a dime.
What luck to be unborn with so much talent;
Never having earned His spot with any effort spent.

Well, we'd still humour our dear parents,
Not telling them where we'd been apparent:
Honoring her offer, on her and off her;
Yet, we'd soon learn what's what, via human nature.

If this non God we'd emulate, we could kill
Those who solicitate—with our free will,
Even time, spouses, bugs, microbes and other swill,
And, of course, outlaws, and, especially, in-laws.

Ah, but the concept of reward and punishment
Handed out by this omnipotent, omniscient God,
Is but derivative of family experience—
The child and parent—a conception of our world.

So, if God's a good role model, a leader,
Someone that we would follow, imitate,
Emulate, be like, adore, or follow,
Then what else would his fine example allow?

We could jail people for the sins of their
Ancestors, exterminate humanity,
Allow known evil to exist and tempt,
And devise devious entrapment plans.

We could have temper tantrums and outbursts,
Envy, or not permit competitors,
Grant free will only it matched our own,
And covet worship, adoration, and praise.

The Christian God is vengeful, demands of,
And tortures us with threats of Hellish shove.
Well, if I were a God and ruled above,
You could remove all my powers but love.

Now, back to the commandments sultry:
Yes, we should surely admit adultery.
But, why banish all thoughts impure,
Those that are simply our human nature?

Now, if He'd wanted us not to be naked, say,
Then surely we'd have been born that way!
As for padding, that would false witness be,
So, please, please keep a breast of reality.

And no loving thy neighbors much too much,
By coveting their Heavenly bodies such,
But, thy own ass do covet—as it's not free;
Follow Moses, by always tying it to a tree.

There are stealers about, another shalt not,
Who take office supplies home a lot,
And take various and sundry restaurant items,
As well as keeping every pen, never buying them.

Now, really, always do one to others, too,
Before they can do the same to you,
And never lie in court; no, not you—
Just let your lawyer do it for you!

Now, walking on water is very much out,
Unless there is solid ice—winter, no doubt,
And never know that sin is fun's evil twin,
And ever enter that evil Sin-a-God.

So, what more would this invented God be,
The One with neither paternity nor maternity?

Would we then be made so specially
That we'd be rewarded for all eternity?

If we'd worship Him from fear of Hell,
Then He'd rightly cast us into it;
If we'd worship Him from a desire for Paradise,
Then he'd deny us entrance into it.

He'd say to Adam and Eve in Eden:
"Do what you like, but don't eat the apple".
Well, we know that when you tell children
Not to touch something, they certainly will!

Only a Fool would blame His own creations
For the flaws therein, for His poor craftsmanship,
So rejoice, there's no Maker of Man—these 'flaws'
Provide for many, interesting character types!

Well, He's still on prozac, so they say,
For He works in mysterious [insane] ways.
The free will to us given is ever free,
Unless it doesn't match His own entirely

So, we'd still think that sins, or ills,
Of a mental nature are caused by the Devil,
An evil tempting spirit; however, now
We know of brain chemistry gone astray.

He'd still detest evil so totally completely,
That he'd allow the Devil to tempt us mercilessly.
And sins, even the most horrible ones, well,
No big deal; we'd just repent them to avoid Hell.

Rigged & jigged, God's perfect plans would be done,
But he'd long for some surprises yet to come,
So He might even roll the dice, it being random;
"Darn!" he'd say, I already knew the outcome!"

One-night stands with engaged young virgins
Would be OK, but those are not good urgin's;
And no fighting, especially if you are weak;
So, when one kisses your ass, turn the other cheek!

Thus, a God-who-is-a-being would, like us,
Be dependent on, and exist after,
The Ground of Ultimate Reality,
And so could not, in Himself, be His own cause.

The Diviner would just sit around,
With nothing else to do,
His mind already full with
What would become as new.
He couldn't play dice,
Scrambling the forecast,

For He would know all
Of which the die was cast.

Now, hail the real All and the One,
Omnipresent, for it's eternal, too,
And can neither be created nor destroyed,
Being its own cause, and the Ground of All—
It is Energy!

SEGNO (SIGN) # 0

"There is the 'vacuum'", replied the other,
"A base state, one pervading all of space,
There being no signposts within it,
Or anywhere, since it is of no direction.

"We must regard it the stuff of which things are made;
For just as all living creatures inhale the air,
So do all the real natures inhale the vacuum."

"This intimation is the mark of manifestation,
A demonstration that's the token of the evidence;
The aetheric and heavenly sign of things to become,
Both the portent and the omen of so much possibility.

"It is both the warning and the present notice,
Presaging both the promise and the threat.
Aft this sign, that the vacuum 'indirects',
Then the real gestures ever beckon;
They of an the unsignal faint,
The wave and gesticulation of you.

"We read the noise of the quantum theater—no marquee;
All is daubed without symbols, to mark no cipher, bare,
No letters, characters, figures, or hieroglyphs there,
No ideogram of the rune of order,
No emblem of the Divine."

**FOR THE SUPER
HEAVYWEIGHT CHAMPIONSHIP
OF THE UNIVERSE:
GOD VS. SCIENCE**

Round 1

In the Beginning...
God played an active role in the Universe,
After creating it, each and every verse,
And especially one upon the Earth...

Which is supposedly
Only a few thousand years old,
Or so it has been told.

God won this round, hands down,
For even those many science clowns
There were there at the time
Thought that man was prime,
Being the special center of creation
And that the sun and the stars, in elation,
Revolved around his nation...

And, furthermore,
That evil spirits caused physical ills
Along with all of our mental ills,
As aggravated by life's frills—
Which were all called 'sins'
That somehow still came from within.

Even fun was one of sin's evil cousins,
For the Bible was made of old Jewish legends.

Thankfully, those hundreds of odd Gods
WHO had come to reign before GOD
Were crushed and by Jehovah trod .

However, about three centuries ago,
The realm of natural law was extended, so,
The Supernatural Kingdom
Began to shrink away, some,

Eventually vanishing from all of existence,
But, we get ahead of our own persistence...

God made Adam fully formed, without a navel;
But, now, an asterisk on page one of the
Philippine Catholic bible says "No",
To not take it literally; it's just not so.

R o u n d 2

God came out quick, still claiming the writ
That he guided the Earth safe through its orbit
Around the the centered sun in space, His Son,
For by now the Earth's motion around the sun
Was known to be true to nearly everyone.

Newton demolished this notion
With his laws of motion.

God thus no longer ruled Nature's course,
For the world was free to run its course.

From Isaac: Laws and Revelations:

There is a mote in space known as Earth,
A pale blue dot of fluff orbiting a hearth...

Due but to Newton's laws of motion, there's none—
No Godly hand guiding it safe around the sun.

The vanishing had now really begun.
The heavens and the Earth were one.

Stars and galaxies went on and on, puffing,
And we became the center of nothing.

God was losing his definition in stone,
As his sworn traits disappeared, one by one.

So, He's retreated to higher ground, that is,
Outside of space, time, and all that exists.

Round 3

God so then claimed to appear to us
Only in spiritual thoughts and ideas, thus,
Making Him responsible, as our Savior,
For the goodness of human behavior.

This metaphor was then found to be unnecessary
As the source of human character non-contrarily;

Yet, some still clung to the life-line ropes
Of His intervention, with their hopes,
Although some claimed that He
Did not involve Himself, or be,
In our daily operations and pleas.

So, God no longer intercedes in causes,
Except in some nebulous cures and "becauses",
As being safe from harm, or curing what hears,
But, He never heals amputees, or appears.

For the latest is that He must stay hidden,
Even if the "miracles" of His Son, bidden,
Were very much out in the open to see;
Better that no one know of Him clearly.

So there is "faith"—a blind trust in the unknown.
Believe it or be tortured—or has this, too,
The Word of God, become inoperable?

Only the supernatural realm remains.

Round 4

God was still yet "seen" to intervene here,
Saving lives, here and there,
In the natural world's reality,
But, too, striking planes from the sky,
Ever adjusting and smoothing the operations
Of natural law, expressing His inscrutable purpose.

However, scientific knowledge, cosmology,
Fundamental physics, chemistry, biology,
Anthropology, and psychology were wont
To undermine religious views on every front.

God was losing His strength to be,
For science loomed large, quite ponderously.

Religious knowledge, without proof,
That which professes absolute truth,
Now fails and fades, an impossibility,
While science, which professes fallibility,
Succeeds and grows stronger daily.

There were still those strange myths...

Why is the Old Testament out of the pew,
In many churches, in favor of the new?

Was it divine revelation or not?

Do God's fits not become a good role model?

Round 5

With God in full retreat, it was yet thought
That at least He had instilled or wrought
A spiritual essence in us willed, whole,
That which was called the "soul".

What vanity to claim such full self-importance!
To demand so much from the universe...

That one would claim an angelic vapor that
Drives a living being, provides character,
Morality, and consciousness, on top of
A burdensome, fragile, and expensive
Organ such as a brain not needing to be used?

Science collapsed the idea of the elan vital
When the synthesis of substance began.

Life's chemistry was of chemicals!

Yet, it was still said that God made all the kinds,
Albeit strangely full of the problematic signs
Of such an unintelligent design,
For how else could it all have been consigned?

Darwin told us how natural selection
Explained the mysteries of evolution
And of the variety of life covering creation...

Extending from animals to us, a continuum,
Now even seen to go back to a bacterium.

We were no longer special at all, as such,
Differing from chimps by not very much.

The discovery of genetics later on
Collaborated it all in our genome.

So, because of evolution's record written
God's Bible was no longer seen as written
In plain text for the common man,
But is open to symbolics and interpretation.

Thus, now, He just is, the same as the universe,
And, yet, this would be a kind of curse,
For this state would be quite restrictive...

Not to mention the mere tautology
Of a universe, a cosmos, and an Entity
Being one and the same pose,
Such as a rose is a rose is a rose.

Since the above Cannot be,
He's now become but a Deity,
Leaving us all on our own,
Our own life to own,
The same as the nonreliance
That is seen by science;
Now we're fully sentient,
But a planned, random accident!

Aye, the truth of what now we are is:
Not made by a Wiz to take a quiz,
But Mammal, organic, of speciation—
One passing narcissism and self-adulation,
Onto the bio-electro-chemical organism
Evolved upon a planet near a star, risen
Of and in the long and winding mindless way
Of slow time, dust, and selection by death
That sifts the best from the rest: evolution's breath.

R o u n d 6

More devastating blows landed, raw,
Einstein's theories extending Newton's laws
To the very large universal scales, with trust,
While quantum mechanics brought us, next,
To the reach of the very tiniest of objects,
There being no place left for us as subjects.

God was nowhere to be seen, having vacated the arena.

Yes, science has found that the universe
Operates just as it would without Him—
That evil spirits don't lead to bad health,
That brain imbalances can lead to sins.

Devil, Hell, the Bible, intercession, etc.,
Are all gone now—he is undefinable—
Protected from the knowing—safe, away:
Yet claimable as the unseeable unknowable!

R o u n d 7

Confirmations were everywhere hatched,
Since scientific laws must ever match
And predict the facts of what it mimics,
For example, of the quantum mechanic.

Although QM's basis seems counterintuitive,
It always works out just perfectly,

For we employ and depend on it, in every way,
On tech products based on it, every day.

Science ever goes on to astronomical heights.

The first supernova since 1572
Appeared in some small galaxies nearby, a few,
Called the Magellanic Clouds, too...

Though its radiation began a while back,
We saw it alight upon us in the 'now'—
Those immerse quantities of energy
Of a mighty star-stuff maelstrom.

A Chilean astronomical technician, some bloke,
Stepped outside, perhaps to have a smoke,
And, being observant, spotted it's yoke!

Ah, he, a mere human standing around
Out under the dark starry sky, aground,
Detected it, upon this lucky time,
For the large telescopes only take in the shine
Of the sky in small sections at a time.

He went in and told of such unexpected,
That a large burst of never-detected
Neutrinos was now to be expected.

The astrophysicists called their colleagues,
C'mon, you all, answer, please,
Those deep beneath the Earth's surface,
In the United States, Japan, and Europe,

And then said, "Look in your tanks, in revelry;
You have already made a great discovery."

They were right on the dime, this time;
Each of the observatories had detected the signs
Of a few tens of neutrinos at about the same time.

Consider the magnitude of this achievement,
For they had tested all of what physics meant!

They had predicted the events that go
On in a star's death throes—
By using theories from nearly every part of physics:
Special and general relativity, quantum mechanics,
Fluid mechanics, thermodynamics, nuclear physics,
Atomic physics, and elementary particles.

If any of these theories had in error flailed,
The prediction of the neutrinos would have failed.

Thus, the laws of nature that are known to us
On Earth everyday must have the same thrust
Hundreds of thousands of light years away;

And, also, the same back in the day
When that star had exploded so,
Hundreds of thousands of years ago.

God had been pushed completely out of the ring,
And so there were no more praises of Him to sing.

There were no immutable forms made,
As is, that never change, as "bade",
For, there was no one miracle of life
Leaping into any living form, but rife
With all of natural selection's strife.

Slightly dead chemicals
Became definitely alive chemicals,
Metabolizing into many particulars,
This being nothing spectacular.

We even have evidence of ancient algae
From 3.5 billion years ago, in a sea,
When liquid water was available and free.

It still took more than two billion years
For more complicated life to appear.

There was no Garden of Eden.

God's become aloof; he's begun to dissolve...
He let the design gradually evolve
Over thirteen billion years into man's plot,
The endless universe a mere backdrop.

He is the Intelligent Designer that
Is deducible from not understanding design,
But, wait, he is of infinite design—
So now I know that something had to make Him!
(Ground—Of—Determination: G.O.D.)

Round 8

The Knockout.

The three-degree blackbody radiation was found,
The CMBR. It comes to us from all around;
Nonuniformities in the radiation were found at last,
Those that formed the galaxies of the past.

The QM realm has been proved, of late,
To be a fundamentally fuzzy state,
Virtual plus and minus states
Popping out at any old rate;

That is, there are no real causes,
For there are no hidden "becauses".

This realm is not quite a Nothing,
But a near 'nothing',
Nor some infinite regress of something.

Virtual particles may take the helm
Or cancel back into the QM realm.

If "Nothingness" were exactly zero, not fizzy,
Then this 'vacuum' would not be vague and fuzzy.

Thus, an absolute Nothing cannot exist to be,
For its very definition means that it cannot be,
As then it could not even be there at all in reality.

So, there is but the quantum jitter;
There was only this 'possibility' forever.

Oftentimes, the QM "virtual" particles magically
Spring into existence, and vanish quickly,
Although they can interact and remain, really.

If not, they have to vanish so quickly
That we cannot account for their reality.

If we could see them, then the QM possibility
Would not be the vacuum fuzzy energy;
But, if they were not there, as something,
The vacuum would be exactly Nothing...

And so this certain school
Would violate the vague and fuzzy rule.

None of these happenings are invisibly lame,
Such as those of the supernatural claims,
For the fuzzy 'nothing' has many effects
That we can compute and detect.

So, is there is no cause, no purpose, unthunked!
Does this make us go into a deep dark funk?

No, for it is our glory that we are free to be,
The making of life being our own responsibility.

Now God was dead, gone, having counted out,
Having never been, whether within or without.

The eternal, causeless ground-state
Could have never had any "create",
For there could be nothing prior
Such as that which is known as a Creator.

Terrorists still go to war in his name;
It's all going astray—this notion fails;
If I knew where the Great Designer stays,
I'd question his mysterious(insane) ways.

What, then, is left of this vanishing Phantom?
More features than I've listed have fallen—
The Extraordinary Superstition's kiss
Remains as but a shadow of a wish.

CONVERSATIONS WITH THE COSMOS

The Universe whispers its secrets
To us from the CMBR, when light was born,
And before, if gravitational waves appear.

X-rays shriek with high pitched terror;
The infrared rains down its stories;
Gravity & dark energy run on through everything.

Cosmic rays pound us,
Even through the shelter of the sky;
Neutrinos slice through us,
Leaving no wound;
Gamma rays tickle us;
Dark matter tugs on us mysteriously.

We need the universal translator!

THE TRADEOFF

Single-celled creatures never die
But live on as their offspring by dividing;
Whereas, via sexual reproduction,
The parents eventually die.

Death is the price we pay for sex.

SPEAKING FOR AND OF GOD

Whatever Austin or anyone makes up
About the nature of God,
Speaking for Him, or even that He is,
Is, in the end, ever made up.

We are not dwelling on this here,
For we are doing the act of 'what if',
But, it's a lesson that shouldn't be lost on anyone.

So, then, deist, I would certainly have to agree
That the Scientific Deity type God
Is the only one that is possible,
Being that He does exactly what nature does.

The supposed 'Theity', is, as known,
But a lot of window dressing
That humans clothed Him in.

And, too, no straightforward beyond, extra-,
Or super kind of evidence can be found
For the Theity who is said to be everywhere,
People's inner sensations
Of their minds not withstanding.

Ah, but the Deity set material in motion
And also constructed its nature
To carry on with His Master Plan,
Which, by the way, was not fixed, like a DVD,
But purposely allowed for variation on the theme.

So, if we may make up more assumptions,
Such as that the Deity has being,
Then the Deity just sits around,
Now that his plan is in motion,
Doing we know not what;
But, there are some questions
That almost give Him pause:

He almost wonders how He got to be there
In His lone position of ruling All that is,
But, then, of course, realizes
That He never 'got there' but was always there.

So, He is not at all worried
That He has no earliest memory,
For there is some magic feature
That allows Him to search infinitely back
At once, into all the happenings previous.

He is almost stumped on why He is there,
As like in "Why Me?", a large complexity,
But, since His 'something'
Is the natural state of affairs,
Rather than the unproductive Nothing,
He almost comes to terms with that 'paradox'.

As must be said, He is so infinitely intelligent
That there can be, of course, no vengeance
Or any bad emotion happening in Him.

Again, humans made Him to be more than He is,
Molding Him in their own limited image.

The only real question, yet, again,
That might give Him full pause
Is how He Himself is already there, fully formed,
In His extreme state of complexity.

Nevertheless, He builds the plan for a universe
That works exactly as designed,
Doing it in less than an instant,
Not in six days, for there is no limit on His power,
Nor does He get tired and have to rest a day,
For His energy is infinite;

So, this Biochemist Scientist
Is the Ground-of-Determination (G-O-D),
Although stripped of all the human holy-moly
Painting of his character and His aims,
Even the lame assumption of His being

A complicated complex composite being
With a magical built-in system
Of a totally functioning mind.

So, when you really get down to it,
After the necessary minimization
To being nature's nature itself,
He is the TOE,
The Theory of Everything.

FOLK 'WISDOM'

What is common sense?
Well, nature doesn't give a hoot about it.

Newton thought his theory of gravity
Was a great absurdity;
Yet it works for NASA.

The moral is:
If it works then it is
What nature intended,
Whether sensible, common,or not,
And not necessarily
The world that we make up.

CURRENT COMPLICATIONS

Life is complicated,
It taking thousands of different types
Of molecules to make a person—
Because it has been
Pieced together by evolution,
Borrowing whatever worked
From whatever ingredients
Were handy to pull off the shelf at the time.

THE FOREVER FIELDS OF REALITY

Michael Faraday introduced
One of the most radical ideas in science.

They thought that he had,
For once, gone too far.

Particles became rather irrelevant,
Being mere spigots through which forces flowed.

The real stuff of reality was the forces flowing,
The particles being only the source.

The burden of reality had shifted,
For the space between particles became primary.

Particles were only the intersection
Of the forces that wove the universe.

Forces create stresses in space,
A superhighway
Of how to get from here to there.

An electron wiggles in the sun,
Tweaking the E/M field;
The ripples travel for 8 minutes
Then tickle an electron in your eye.

You see the light;
Light is a tweak.

Physics has never been the same since.

The field concept became real,
The idea being the same as the thing,
Fudging forever the difference
Between something and nothing;

Yet, fields are made of something real,
For they have energy.

Einstein called the field that be
"A change in the concept of reality...
The most profound and fruitful one
That has come to physics since Newton."

Matter, then, is simply a place where
Some of the field happens to be concentrated.

Matter travels like a wave in a rope,
But, the rope itself does not travel.

The field is not so much
Something in space,
But more like of space.

This is why all particles of a type are identical;
For they are each manifestations
Of their fields everywhere the same.

The field takes on a life of its own,
Even when the object that created it is gone.
The traveling kinks continue;
They propagate endlessly.

Where the vacuum is free of matter
It is not free of field, but filled with it.

Energy and mass are the same stuff,
But it takes a whole lot of energy to make mass.

Field is thus the bridge
Between matter and empty space.

Fields can't go away,
As they're part of the structure of the vacuum;
When in their quietest possible state
They are the vacuum.

This is about as close to nothing
As anything ever gets.

Forces act on things,
While matter is acted upon;
You can walk through a field,
But you cannot walk through a wall.

Kinks in fields can pile atop one another;
Kinks in matter hold each other at arm's length.

Yet, somehow, beneath it all,
They are kindred spirits.

Faraday made fields real;
Quantum mechanics made them magic—
And lumpy—the currency of QM.

Everything melts, via uncertainty,
As when we try to measure a quantum property.

But this, too, means that no quantum property
Can ever be zero, for zero is a precise amount,
That is, it is that motion can never cease.

Try to pin down an electron,
Such as putting it in a box,
And it increasingly moves about, ever faster.

It is heads or tails while it is still spinning?
Well, it is just a fuzzy 'both' yet neither.

In a way, QM eliminated
The very idea of zero
From the physical world,
As 'nothing' never sleeps,
But is ever up to something.

(The Loan Shark)

An unusual track was found in a cloud chamber
That Carl Anderson was using
To watch the trajectories of cosmic rays
Streaming in from space.

The track was like that of an electron—
Except that it curved backwards
Under the influence of a magnetic field.
It was the positron, now used in the PET.

A particle and its antiparticle annihilate,
Giving back, in the process, the energy it took
To create them in the first place.

Do they live on borrowed time and energy,
A creation near 'ex nihilo' all over the universe?

Can they sneak out of the vacuum
So long as they snuck back in again
Before you noticed?

"What is the point?"
Thought Richard Fenyman:
"Created and annihilated,
Annihilated and created—
What a waste of time."

They come and go like dreams,
The lighter ones, like electrons,
Popping out more often.

They are the ghosts of the yet unborn.
The road from 'nothing' to something
Goes in both directions;
With enough energy
They can become real.

The so-called 'vacuum' is creative.
The field fluctuates this way and that,
But, on average, the net energy is 'zero'.

The once melted vacuum fell and froze,
Gaining structure,
Such as when water becomes ice.

NATURE'S COMMANDS

1.

Nothing was ever created, that is,
Meaning the ultimate underlying basis,
Because it couldn't be made from Nothing,
For, Nothing, a lack of anything,
Has no 'where' nor 'when'
Nor any properties to be productive.

Not that there couldn't have been
A total lack of anything,
But, that was not the case,
And if it were it would still be the case.

If 'nothing' had some capability to divide
Into plus and minus
It would not have been Nothing
And it still would have
Always been there always.

2.

The ultimate underlying basis
Was the natural state of affairs, 'forever';
There was no creation;
As such, there was no creator or Creator.

3.

Nature ever proceeds from the simple
To the more complex and composite:
It goes to quarks and electrons to protons
To stars to the lower atoms being emitted
To the higher atoms from supernovae
To molecules to cells to life
To mind to consciousness.

4.

Of how nature works, in every way,
Nature may eventually tell.

It is that we may find out or not,

But, it not to be found anyway else,
For Nature is the embodiment and result
Of what ever was and still is
In its various arrangements
Of complex composites.

5.

Thus, we are free to be, within our form;
Plus, existence must be dealt with
First and foremost, over essence.

SOUL FOOD

The soul lacked of experience and love
Having nothing with which to think of;
For no memories of sensations could it pave,
Nor any new associations could it save.

All was non-sense, such as in meditation,
When there ar no thoughts or perception.
This nothing is such boring bliss,
So I'll see what it is to exist.

So the souls looked for babies born
And latched thereupon their horn.

WINDINGS

I heard that atoms
Are of energy spun
And condensed
Into solid pegs
To hang a universe on.

9 STEPS BEYOND

1.
What is seen is what Nature has wrought.
To ignore what is known,
Such as having the brain not do everything,
Is a first step 'beyond' reality.

The 'expensive' brain uses
Almost half of one's energy;
Even the visual system is a part of it,
As well as every nerve spindle in the body.

The brain does both the subconscious
And the conscious work;
Thus, there's no place left for anything else.

2.
Stating that there is a 'soul'
Is both to ignore what is
And to make up what isn't.

Now we see that's why
The brain had to go in (1);
It's that the direction of the brain
Would have conflicted with that of the soul.

This ignoring and inventing
Is also accomplished when the item of (1)
Can't really be minimized,
Such as that the "ego"
Was not meant to be used,
That Nature got it wrong.

Both stem from the (in)vested interest in (4).

3. So, the soul is declared invisible,
Yet another step beyond. Why? See (4).

4.
The soul was placed by God.
Another step beyond.

5.
Going yet beyond,
Further structures are then layered
Upon the vested interest of having "God".

6.
The God theory is then held as fact,
Rather than a could-be,
Yet another step beyond,
Now becoming unethical—
A deception, for God is just a theory.

7.
The God theory is then preached as truth
To impressionable young minds,
Who, naturally, tend to believe
The adults' merciless indoctrination—
A step way too far since innocents are affected.

8.
The flawed concept of religious 'good' leads
To its inverse as being 'bad'
Or even when different from
The other flawed religious 'goods'
And so is labeled as 'evil', in either case;
The 'goods' must be defended;
Their credibility ever lessening
Due to the others existing.

Anger. Division. Crusades
War. A mess.
The steps afar are due to
The initial invested interest
In the invisible imaginary
For which no evidence
Is even conceivable,
By definition,
For the belief hangs in the air,
Ungrounded.

9.
And yet,
The ultimate and simple underlying basis of All,
That which went up onward to the composite
And to our own complexity, was never created,
Having to have always been there
Since it could not have been created from Nothing,
For Nothing cannot 'be' or 'do'.

The eternal original, tiny, simple 'something',
Ever there,
Means that there was no creation or Creator.

Plus, complexity falls at the other end of the spectrum.
They were looking in the complete wrong direction!

STABILITY MARCHES UPWARDS

Everything
Becomes less complicated and smaller
As we delve downward,
From the complex toward the simpler,
As it must;

And so then does the simplicity of
The Theory of Everything,
For complexity lies
At the other end of the spectrum,
Where we are.

EVOLUTION

A monkey in a silk suit is a human being,
Which doesn't fit or sit well with the seeing;
But the case has been more than proven;
We are animals yet, crusin' for a brusin'.

THE ILLUMINATION OF SCIENCE

The Illuminati of today
Are not those of the past,
For they have mutated,
Some even picking up on notions
That others falsely ascribed to them.

Back then they were scientists.

The Church and its scripture
Is basically immutable,
Still according to the myths of old;
That is, the idea of Jehovah told
Wiped out the numerous Gods of old
And became the new—
The one and only, too.

Religion may not be burning scientists
At the stake anymore, but if one thinks
They've released their reign over science,
One must ask why half the schools in the U.S.
Are not allowed to teach evolution,
Why the U.S. Christian Coalition
Is the most influential lobby
Against scientific progress in the world...

As for the Illuminati of old,
The obliteration of Catholicism
Was their central covenant.

The brotherhood held that
The superstitious dogma
Spewed forth by the church
Was mankind's greatest enemy.

They feared that if religion continued
To promote pious myth as absolute fact,
Scientific progress would halt,
And mankind would be doomed
To an ignorant future
Of senseless holy wars.

And, I might add to the above,
Much like we see today.

So, Bush killed stem cell research
And went to war against Iraq
After consulting with a 'Higher Father';
Holy wars now being everywhere, since
How could the other religions be so wrong!

Those trying to hold a monopoly on truth
Cannot help but to label the contrary as evil,
And, thus, act accordingly.

So, we do have to worry, still,
When the Church wants to be
The sole interpreter of the 'truth'.

Flawed and arbitrary concepts of good and truth
Only cause the contrary to be labeled as evil.

Ah, thought Galileo,
As he wandered past the deserted
And flower-grown ruins of Rome, one night,
*This looks to be the same now as it will and was
A thousand years before and after me.*

*Would that there could be a day
When science was free,
When the once great Roman glory
Would pale beside that brightest light of day!*

Galileo looked about and around and behind;
No one was following him to his ultra secret lair,
Where other scientists would join him again
On this starry night, safe therein to congregate
And discuss the topics forbidden by the Vatican.

(To this day no one has found Galileo's lair,
Called The Church of Illumination.
I am obtaining all this information about Galileo
From his little known 'lost' diary.

I even have an unpublished book of the Holy Bible
And a few of Leonardo's 'missing' diaries,
But, those are other stories.)

...go to Rome, which is the sepulchre,
Oh, not of him, but of our joy: 'tis nought
That ages, empires and religions there
Lie buried in the ravage they have wrought;
For such as he can lend,--they borrow not
Glory from those who made the world their prey;
And he is gathered to the kings of thought
Who waged contention with their time's decay,
And of the past are all that cannot pass away.
(Shelley)

Galileo noted the ancient sculptures
Still standing against mouldering time,
Knowing that the new scientists arriving,
If they were worthily smart enough,
Would have to use the clues provided
As the way to the secret meeting place,
For there was no map made and never would be.

As the word of this
Scientific brotherhood began to spread,
Scientists would travel thousands of miles
But upon the slim hope of chancing a glance
Through Galileo's fine telescope
And discussing the master's many ideas.

Go thou to Rome,--at once the Paradise,
The grave, the city, and the wilderness;
And where its wrecks like shattered mountains rise,
And flowering weeds, and fragrant copses dress
The bones of Desolation's nakedness
Pass, till the spirit of the spot shall lead
Thy footsteps to a slope of green access
Where, like an infant's smile, over the dead
A light of laughing flowers along the grass is spread;
(Shelley)

As Galileo wandered among the ruins
Made one with Nature in their decay,
Or gazed on the Praxitelean shapes
That thronged the Capitol,
And the palaces of Rome,
His minding soul imbibed all the forms,
This loveliness becoming a portion of himself,
As well as its science, even right here,
Within the realm of the Pope's Holiness
That shadowed him—
Much as the darkness of night
Condemned the day.

And gray walls moulder round, on which dull Time
Feeds, like slow fire upon a hoary brand;
And one keen pyramid with wedge sublime,
Pavilioning the dust of him who planned
This refuge for his memory, doth stand
Like flame transformed to marble; and beneath,
A field is spread, on which a newer band
Have pitched in Heaven's smile their camp of death,
Welcoming him we lose with scarce extinguished breath.
(Shelley)

Many had been burned before, thought Galileo,
So 'tis a difficult path to follow,
Yet the truth calls me forward...
And so he had published
The 'Starry Messenger'.

Later on, Galileo had argued
That the Bible had to be interpreted
In the light of what science had shown to be true.

Galileo had several opponents
And they made sure that a copy of
The 'Letter to Castelli'
Was sent to the Inquisition in Rome.

In 1616 Galileo wrote
The 'Letter to the Grand Duchess'
Which vigorously attacked the followers of Aristotle.

In this work, which he addressed
To the Grand Duchess Christina of Lorraine,
He argued strongly for a non-literal interpretation
Of Holy Scripture when the literal interpretation
Would contradict facts about the physical world
Proved by mathematical science.

... Galileo walked on slowly,
For his health had become poor,
And noted the setting moon—
The sky would be wonderfully dark.

He would soon be found guilty and condemned,
But he knew none of that this night.

The eventual 'Father of Science'
Again sat with the scientific Illuminati of his time,
The discussions as free and glorious as ever...

He was later put under house arrest
In his home in Florence,
Having by then nearly gone blind,
But the starry memories of the Milky Way,
The moons of Jupiter and more
Remained in a mind still free—
That which could never be taken away by 'Dogma'.

His body was concealed
And only placed in a fine tomb
In the church in 1737 by the civil authorities,
Against the wishes of many in the Church.

On 31 October 1992,
350 years after Galileo's death,
Pope John Paul II gave an address
On behalf of the Catholic Church
In which he admitted that
Errors had been made
By the theological advisors
In the case of Galileo.

He declared the Galileo case closed,
But he did not admit that the Church was wrong
To convict Galileo on a charge of heresy
Because of his belief that the Earth
Rotates round the sun.

The Torch Passes Its Light

His eyes were so weak
"That he could no longer see the sky."

A young Illuminatus embarked on a long pilgrimage,
*"A sojourn to Galileo's delightful villa at Arcetri,
Just beyond the walls of Florence.*

*"There it was that I found and visited
The famous Galileo grown old,
A prisoner to the Inquisition,
For thinking in Astronomy otherwise
Than the Franciscan and Dominican licensers.*

*"I was his last disciple, as you say
I went to him, at seventeen years of age,
And offered him my hands and eyes to use."*

Galileo recalls the momentous occasion
("that day of days"):

*When, quietly as a messenger from heaven,
Moving unseen, through his own purer realm,
Among the shadows of our mortal world,
A young man, with a strange light on his face
Knocked at the door of my house.*

His name was John Milton.

Milton at the gate: Friend! let me pass.
Dominican: Whither? To whom?
Milton: Into the prison; to Galileo Galilei.
To this, the Dominican guard protests that,
Where Galileo is being held, there are no prisons,

Only confinements of sorts
For those guilty of "heretical pravity"
And "other less atrocious crimes".

Not to be taken in by such rhetoric,
Milton stands his ground and demands
(on divine authority)
That the gates that confine the great astronomer
Be opened at once.

Responding to the demand,
The Dominican guard
Can only admire the young man
Who confronts him.

To himself the guard exclaims:
"What sweetness! what authority!
What a form! what an attitude! what a voice!"
After which he acknowledges
That his "sight staggers; the walls shake;
He must be—do angels ever come hither?"

...Plots had been perhaps laid against Milton
As one who had 'seen' and 'heard'
Matters that were best left untold.

In Galileo, 'frail and old,' Milton had 'seen'
One of those near blind illustrious
Of whom he had so often dreamt,
And of whom he was to be himself another.

O, dark, dark, dark, amid the blaze of noon,
Irrecoverably dark.

Some thought that
Milton's Lucifer (Latin for 'light bringer'),
Came off much better in 'Paradise Lost'
Than did God Himself.

Lieber in der Hölle regieren als im Himmel dienen.
[Better to reign in Hell, than serve in Heaven.]

'Twas here, his final resting place,
In a church...
At last enshrined
As the Father of Science.

Embellished, as the Master in stone,
He's ever looking up
Whence forth came the light
From the starry skies.

The Fanciful View From Today and a Review

Back when religion persecuted science,
The Illuminati became a secret organization
Taking refuge from the scourge of the Church.

The Path of Illumination

In 1600 Rome, the Baroque theatre
Of political intrigues and inquisition trials,
One of the most influential secret societies
In history was born: The Illuminati.

Gian Lorenzo Bernini and Galileo Galilei,
The twin heads of the society,
Scattered throughout the Eternal City
Clues and enigmas which, once solved,
Would lead Illuminati adepts to a hidden lair.

It was thought that the rumored ambigram
For 'Illuminati' could never be found,
It reading the same upside down.

There were Four Altars of Science,
Representing the four elements
Of earth, air, fire and water,
And a mysterious text from John Milton,
They being the key clues that, once decoded,
Would lead on the Path of Illumination.

DOOM?

"Behold this droplet of anti-world,
My anti-matter that LHC created,
Enough material to see."

"My God, a visible amount!"

"See, here it is, suspended
In a vacuum in this tube,
For even the air would ignite it."

"Quick, send it away,
Get rid of it."

"No, for I have discovered Creation."

OIL AND A LUBE

I changed my mind so it would work better:
Cleaned out the clutter of some old letters,
Removed the cobwebs, that all unfettered,
Plus that to-do list, from those go-getters.

THANK YOU, CELLS,
FOR BEING WONDERFUL

Every living thing is a wonder
Of tiny atomic engineering;
So, let us hail and bow to our cells,
For they do everything!

THE MEADOWS OF HEAVEN

We, of the highest consciousness ever known
And of the most versatile form that's been shown,
Reside as consequent beings in this Earthly realm;
Possibly the most fortuitous creatures
That the universe has ever wrought.

In fact,
We are this universe come to life—
Necessarily from a long line
Of fortunate accidents.

It had to be this way, for any universe
In which we could emerge
Would have to be appropriate for us
Or we wouldn't be here to discuss it.

Looking back,
We already know, ahead of time,
That we will discover
The many 'happenings'
That made us possible.

All this we know and expect
Because we are here.

Perhaps, in some other 'wheres',
Junkyard universes litter the omniscape,
For they flunked, failed, and miscarried—
A quadrillion trillion universes broken down
For every one that worked to any extent at all.

In some of these forlorn universes,
Perhaps the material was inert
And so it just sat there, doing nothing, forever.

In others, maybe gravity was insufficient
Or had no natural place to collect particles,
And so it thinned out endlessly,
Spreading coldly toward infinity.

In yet others, again,
Even those in the same ballpark as ours,
Perhaps the portions weren't quite right;
Although they may have formed a few elements,
They went no further than that for a zillion years.

In our universe, the dark chest of wonders
Of Possibility and Probability opened up
In just the just right way:

Naked quarks spewed forth,
Among other things,
And boiled and brewed
In one of the steamiest broths
Ever cooked up.

They somehow simmered and combined
Into the ordinary matter
Of protons and neutrons.

Then, quite independently,
By some unknown means,
Dark matter/energy arose, as well,
In just the right mix, and, luckily, too,
Some very long filaments,
Called cosmic strings,
Formed and survived long enough
To be useful as collection agents,
Which were merely imperfections,
As in an unevenly freezing pond—
A kind of a cooling flaw.

None of these happenings
Were related or connected,
Except by Potential's destiny,
So, 'fortunately',
The cosmic strings attracted,
By their gravity,
Both dark and ordinary matter,
Which, in turn,
Attracted even more of the same.

These pearls of embryonic galaxies arose
And were strung along these cosmic necklaces,
As can still be noted today.

So it was
That some almost incidental irregularities,
Frozen out as cosmic anchors,
Were latched onto by matter, both light and dark,
The proportionate portions of which were favorable,
The dark matter dwarfing our ordinary matter
For some reason of a happy 'circumstance'.

Fortuitously, as well,
Anti-matter, if there ever was any,
Did not fully cancel out the uncle-matter.

The universe could not foresee any of this
In and of itself's fundamental substance(s),
For, if it could have,
Then we'd only have the larger problem
Of how the foreseer could have been foreseen,
Ad infinitum...

So, it could have been like the 'trying out'
Of all possibilities in superposition...
A brute force happening of every path gone down.

We know much of the rest of the story
Of how the stars and their supernovae
Created the light and heavy elements
Which combined into molecules,
Which, auspiciously,
Became able to replicate themselves, as DNA,
And progress to make cells, tissues, and life.

And then there was the luck of oxygen,
A mere waste product of photosynthesis
By bacteria, and later, plants,
That could fill the lungs,
As well as build an ozone layer of protection
From the harmful rays of outer space.

Luck on top of luck, good fortune,
And then prosperity...
'Stumbled along' the right path.

Of course, all this took many billions of years—
And it is, of course, this long 'yardstick' that
Baffles the mind and sticks in the throat,
But demonstrates the long time lag needed
To produce even the tiniest of advances.

It bears all the hallmarks
Of randomness at work,
Although quite probable
If Potential had it all 'worked out'.

Dinosaurs roamed the Earth
For over two hundred million years—
Imagine the length of that time.

They were supreme and invincible—
The kings of all the Earth 'forever',
On land, sea, and even in the air—
Heading towards forevermore and beyond,
But...

Dame Fortune once again intervened
When the asteroids, or some such catastrophe,
Finished off the dinosaurs,
As well as 90% of the existing species.

This random event left a vacuum
In which newer species could thrive.

Proto-man gave way to near-man
And thence to us, eventually,
When two 'monkey' chromosomes fused together,
Making 'us' incompatible with the chimps

And so our ancestors, then,
Truly descended from the trees!

We came to need no specialized niches,
Since we could adapt to any terrain,
Having brains that could learn much more
After birth than instinct could bestow before.

Our higher consciousness
Was the crowning glory—
We had won the human race—
The be all and end all; the grand prize
Of the universal lottery.

So, there is nothing more,
Aside from our own progress
To be and learn.
This is it!

DNA remembers every step of our evolution—
And you can see this in 'fast' motion
When embryos form simply in the liquid womb,
Replicate, and then grow cells
That diversify into a human being
After going through some nonhuman stages.

Thus, four billion years compresses into
The nine months of pregnancy.

So, then, hail, and good fortune,
Fine fellows and ladies,
And welcome all of you
To the Meadows of Heaven—
The highest point of all being,
Although we are surely still in our infancy.

The path "chosen" by Potential ends here,
With our consciousness
For that may be what actualized reality.
(Fanciful, indeed)

Rather, there were many pockets of universes,
And is was this very one that could sing our verses.

The further design
And the role of mankind
Is now in our hands.

We were borne here upon the shoulders
Of so many who have long since come and gone,
All of them advancing the cause,
Over eons of wiles—so here we are.

Fare thee always well, fine friends,
For we are some of
The luckiest sons and daughters of being
In a rare universe well done.

Celebrate; live; be,
For everyone dies,
But not everyone lives.

SOLAR RESEARCH

Some researchers landed on the sun and didn't get burned,
Although it wasn't because the anti-light suit they learned,
And it wasn't through mind control that dimmed the light—
It was because they went at night!

WHO/WHAT DUNNIT?

Who or what put fundamental substance here,
Or if it were always, who/what the arena?
How and why the amount of substance present?
From what were its size and forces specified?
Was it a Somebody or a happenstance?

THE GLORY, OR NOT, OF LIFE

Even if some squander their existence,
It is still on offer for the remainder—
Ever those who advance the human race,
Albeit it still in its childhood.

Those who are evil
May have either been
Brainwashed by dogma
Or have various mental defects
Beyond their control,
For brains are far from perfect.

The central nervous system
Still sends out
Thousands-of-years-old messages,
Such as flight or fight;
For, our existence,
In these more modern times,
Has been that
Of a mere evolutionary instant.

Too little serotonin,
Due to genetics, wrong food,
Or lack of exercise
Places the older,
More primitive portion
Of the brain in control.
People then fall for
Their very own thoughts,
Hook, line and sinker,
As gospel, and then behave poorly.

Some realize this later;
Some never do, nor can they,
For all persons
Are not created or raised equally.

Too much dopamine
Causes endless novelty seeking
And risky behavior.

A low standing heart rate
Reduces the anxiety
That should come with criminal behavior
(Usually combined with the other two conditions).

As part and parcel of the organic world,
We are necessarily bio-electro-chemical beings
Born of a nature that knew neither good nor bad
Except how it could go on for continued survival.

Now we know more.

We survived, partly,
Because of our violent abilities,
Not in spite of them,
And partly,
Through the cooperation
For that same hunt or war.

All such tendencies mix together now.

Fear can be turned inside out
(Of public speaking, say)
Into the excitement of the event;
The beast in us can lead to a zest for life.

Certainly, though, drug users, drinkers,
Gamblers, sports nuts, grouches, war mongers,
Quarrelers, users, and all those bad types
Eventually come to ruin—
Some even to regret and/or change.

We could say that they should know better,
That they ought to have, that they can;
But, the disproof of this wishful thinking
Is that so many don't, won't, and can't.

Learning the bad stuff
Can be a dangerous thing,
For learning actually rewires the brain,
Grooving it.

Many never forget how to ride a bike,
Or, unfortunately, how to commit a crime.

Thoughts turn into words,
Thence into actions,
And finally into one's character,
There being a whole spectrum of behavior.

Such is the price of the freedom and opportunity
That is presented by being human.

TO BE

Energy always had to be, perpetually
As Nothing could not freeze in place,
It not even being there.

DEVELOPING NEGATIVES

Words, those written on water of the past,
Emotionally and subjectively cast,
Being of such a specific mast,
Are without a meaning vast,
As, hardly, before they can even pass,
They fade and die so very fast;
For, it's only the universals that last.

ONE MILLION YEARS AGO

The ancestors of the Denisov hominid
Left Africa, a third species
Living among "us" and the Neanderthals.

A specimen, "X-woman", actually a child,
Lived in the Siberian caves
40,000 years ago.

STARRY NIGHTS

Above me, fires burn the stars away;
Below me, the Earth turns under my feet;
Within me, unworded dreams haunt my soul;
Around me, night pours blackness on the ground.

Yet, inspiration returns with the stars—
A thousand ideas beckon from afar;
Ideas wink like fireflies on the mind's meadow—
As starlight, they stab the darkness of naught.

The stars' light is the origin of our being,
The source of our matter, energy—everything;
Permanent, reassuring, and unquenchable,
It's our radiant soul, our self-winding mainspring.

Soul to soul, it said to me, I'm the light,
Thy spirit's sight, a beauty bold and bright,
An inspiration come from darkest night;
I'm a newborn star aglow with insight.

Oh thee, of thine, whence came this life of mine?
I wish thee to thank for this living wine.
Oh Nature, Father Time, Guiding Star—
Thanks for throwing me an earthly lifeline.

Look at the stars in the depths of the night;
Hold the flames in your mind, keeping them bright.
Their power flows, energizing you from
The Eternal Charger—you see the light!

Stars generate the lower elements;
Supernovae generate the higher ones.

Atoms form the molecules that lead to
Life's complexity—from simplicity.

The stars are eternity's running-lights—
They shine, even through the fathomless night!
From what bright star came the gleam in your eyes?
To what distant sun returns your smile's light?

Born of stardust and nourished by sunlight,
I fill my cup with wonders of delight.
Life is a treasure, a radiant gem,
A vision that I'll never see again.

From Heaven's stars came our dust eterne;
Time's seas nurtured thee and thine in turn.
From time, death, and dust we thus became,
And by this, thus, and that we must return.

Purgatory's on Venus, where sulfurs rain.
Hell's found in the sun's heart, oh, hot burning pain!
Of Heaven's site, no one has any idea—
It's the world's best kept secret: Earth's its name!

Earth's a garden, an oasis in space,
A world of boundless beauty and grace.
One could search the heavens for such in vain,
Finding no equal, anytime or anyplace.

Elves find Venus shining in broad daylight,
Knowing where to look as if it were night,
Then follow her as the evening star,
Till with her fiery lover she takes flight.

Just before dawn, amid the dew and moss,
Elves ride on a moonbeam made of Bugloss,
And see the North Star and the Southern Cross
In the same sky, 'most all the way across.

The sun fills the waking and breathing world
With the fire of her imagination.
In poetry, the sun is the power behind the mind;
The moon, planets, and stars are symbols, too.

Sometimes intellectual beauty is bright
And ideas gush from the eternal flame;
Sometimes it fails when the shadows of clouds
Dim the clarity of thought now and then.

Quenchless, boundless, ever bright and burning,
The mind's light searches every dark cavern,
Probing, imagining—its beam alighting
Upon the earth or high atop cloud mist,

And melts, with heat, energy, and desire,
The fog of lone reason and pure passion,
Burning it away, soft dissolving it
With the love of life, earth, mankind, and star—

From which comes adventure, friendship, delight,
Joy, success, triumph, and lasting gladness
Throughout the sun's journey into the night,
When stars shine on mind—suns they also are!

The moon fills the sleeping and breathing world
With the icy coolness of chaste reason
Unaffected by deep burning passions,
Although sunlit to glow in its wan light.

Reason, unsteady as the variant moon,
Oft does not rise in the night to guide us,
And deserts us in darkest times of woe;
We are alone on a black cloud-bound night!

Else the moon hides in the bright light of day,
Or is lost behind an overcast sky;
But, moonless nights take us beyond reason
When the stars excite us with their lights.

Yes, inspiration returns with the stars—
A thousand ideas beckon from afar;
Ideas wink like fireflies on the mind's meadow—
As starlight they stab the darkness of nought,

Until star-like Venus rises near dawn.
Goddess of romantic love and passion,
She captures us within emotion's swell,
While comets flash and confuse the wild sky.

Soon intellectual beauty returns,
Borne on birds' wings as song into the dawn,
For, all human music is but a part
Of earth's ancient melody and rhythm.

Imagination now soars past a day,
And into the season of spring's fast growth;
The shade is deep and cool, like the ghost of
Winter passing—gone but still remembered.

Earth's a garden, an oasis in space,
A world of boundless beauty and grace—
One could search the heavens for such in vain,
Finding no equal, anytime or anyplace.

Why is the Earth for human life so perfect?
And billions of other planets so unfit?
Well, if this world wasn't right for life, then
We wouldn't be here to ask about it!

We are life's eternal creative smile,
Beaming as the universal epistyle.
In us the cosmos has come alive;
Thus we borrow life from Death for awhile.

The void pulsates in a structured sequence.
A field is present throughout space immense,
Out of which all particles must condense—
Occurring where the field's extremely intense.

Atoms are just bundles of inertia,
Knots in the field and fabric of space;
Yet matter defines the structure of space...
The Yin is in the Yang, and vice-versa!

At Heaven's birth, positive energy
Became matter—countered by gravity,
Whose attractive embrace was negative;
So, still, their sum adds up to nullity.

Plus and minus from "nothing" came to be,
But while most charges rejoined, some went free,
The pluses forming matter, energy,
And the minuses forming gravity.

Thus from "nothing" was written our account,
And to nothing we'll still have to amount;
But, in between those two parentheses
The pluses rain on us from Heaven's fount.

The stars' light is the origin of our being,
The source of our matter, energy—everything;
Permanent, reassuring, and unquenchable,
It's our radiant soul, our self-winding mainspring.

Life's a continual cosmic energy dance,
From some ultimate underlying happenstance.
We're immersed in matter's universal rhythm;
Therefore, we must all participate in the dance.

THE GOOD EARTH

Earth couldn't be farther out in space, alone;
In all directions it rolls along, unknown;
Look at the stars piercing the depths of time
They beckon, warm and welcome, the fires of home.

Oft I drink-in the pleasures of creation,
For what else could be the point of cognition,
If not to absorb all that comes streaming in?
Life's sensation is the main attraction!

Earth's a garden, an oasis in space,
A world of boundless beauty and grace
We could search the heavens for such in vain,
Finding no equal, any time or any place.

WHY ASK WHY?

All the stars roll by for me to classify
Science more and more my life does simplify;
But, I have one final question left to ask:
Why in the world was I born to live and die?

Since life's complex, they say it must have origin—
It couldn't have made itself or always have been!
The answer: God; but, they've begged the question
God couldn't have made himself or always have been!

A thousand starry goblets fill the sky,
So we can taste Heaven's drink when we die.
This is man's tale, not God's, so drink today—
The stars shine on, heedless of where we lie.

When I chased the flitting shadows of some
Unknown ultimate perfectionate ONE,
The phantom fled at my touch, a dim image—
Reflected faint and far removed from.

Knowing that I can't solve the eternal mystery
Frees me from that senseless task and all its misery.
Now I see, hear, smell, feel, and drink into my being
ALL reality that penetrates sensibility.

Why fret about life's ultimate secret,
For whose thoughts can escape this worldly net?
It's so easy: don't despair, be happy!
All told, 'tis best to live without regret.

Helpful effort, or love, defines what's good;
Goodness taken to extreme is called God.
Laziness, or non-love, is but neutral.
Evil, or harm excess, names the Devil.

Good and evil—you can't have one without
The other; so, too, with plenty and drought,
Sadness and smile, life and death, night and day,
Sun and flood, give and take, and truth and doubt.

In Heaven, desired pleasures will come like rain,
Or so we've dreamed till we felt no mortal pain;
But we needn't wait for some promise beyond,
Since on Earth—enjoying life—we have the same!

If we were angels, life would be so just;
Instead, we try, we push, we climb, we lust,
We dance, we dream, we feel, we love with zest—
Yes, all this, thanks to the beast within us!

Purgatory is on Venus—where sulfurs rain.
Hell is found in the sun's heart—oh, hot burning pain!
Of Heaven's site, no one has any idea—
For it's the world's best kept secret: Earth is its name!

In the darkness I alit from the Wiz,
Then tried to make sense of this world of His.
Now I've found the answer to life's dark quiz:
One must live this life by what light there is.

DEATH, DUST, AND TRANSFORMATION

All that we knew, even the loveliest and the best,
Decomposes into the dust of earth compressed.
Those songs we once composed now lie in repose;
With this dust the future may arrange and recompose.

THE ADAPTABLE US

Once 'we' hated oxygen, but we now breath it free;
Once 'we' lived underground and later in the trees,
But, now we love it right here, walking on the ground.
'We' once had fins, laid eggs, grew fur; how's that sound?

RIGGED

God longed for some surprises to come,
So He rolled the dice, this being random;
"Darn! I already knew the outcome!"

GRATITUDE

Oh thee, of thine, whence came this life of mine?
I wish thee to thank for this living wine.
Oh Nature, Father Time, Guiding Star—
Thanks for throwing me an earthly lifeline.

Love's spirit wove my soul's warp, weft, and wave,
Creating an eternal, perfect braid
Wound from strands of Truth, Goodness, and Beauty;
Each different forms, but from the same ALL made.

Luckily, I live at peak atop life's pile
Of miraculous lives through eons of wiles.
I'm alive, thanks to all who have come before—
How could I live by any style but to smile?

Pretend that you're dead and gone, siting on a star,
Regretting there the emptiness of life's memoir,
Thinking, *If only I could live it all again!*
Fancy yourself not here, then smile because you are!

Say Farewell! Heaven's promise is bereft.
Still, live with gratitude—be not distressed;
Yet, dismiss immortality's dream;
Accept, with appetite, whatever's left.

I've said Good-bye to the dream of forever;
I'm too philosophical to be bitter.
Poignantly resigned, I accept, with hunger
And joy, all that's left—whatever—with pleasure.

'Twas a time before birth when you were not;
'Twill be a time again when you are not.
From Death your life was a borrowed debit;
Spend it, love it, and live it to your credit.

PHYSICS

Does Atlas underlie the universe,
Standing upon the back of a turtle?
Is there an eternal basic substance—
Or is it turtles all the way down?

Energy is eternal, for it can
Neither be created nor destroyed,
Being made but of itself. Omnipresent,
It's the Mother of all Reality.

The Universe is the ultimate free lunch,
It bubbling out of no-where into now-here
From the quantum foam, via strings or quarks,
The Ground of Ultimate Reality.

Time, too, has a shape in 4D space-time,
And thus, like space, could not have existed
Before the birth of the universe, so,
Our creation didn't wait forever

I'm the All and the One, omnipresent,
For I'm eternal and can neither be
Created nor destroyed, being my own cause
And the Ground of All—I am Energy.

Being is to doing as ground is to figure
As subject is to object, as essence is to existence,
As Awareness-Consciousness is to mind-brain,
As the ultimate simplicity is to the composite.

The universe bubbled out of nothing,
Pluses forming matter; minuses, forces,
All in perfect balance, self-sufficient,
Needing nothing outside of itself—zilch.

MOOD REGULATION

Life's still emotionally primitive:
Negative feedback mechanisms in
The central nervous system, now useless,
Send out thousands-of-years-old messages.

Emotions are molecular events,
Some forced upon us all, like jealousy,
And others, like aggression, born from low
Serotonin, but NOT from the Devil.

Low serotonin stems from genetics,
Stress, lack of exercise, or the wrong food,
And can cause anger, anxiety, and
Depression, even bad behaviors, crimes.

Since the aggressive urges leading to sins
Are not caused by (D)evil, the sinners are not
To blame, although we still have to lock up
The violent ones to protect ourselves.

Prolonged coercive job stress damages
The immune system and brain transmitters,
And causes secondary time-stresses,
Sleep deprivation, and social problems.

Hostility stems from low serotonin,
As one falls for moods hook, line, and sinker,
Kicking the cat or kids when one gets home,
Rationalizing that they made some noise.

Alcohol can raise serotonin in
The short run, but decreases it long term—
A doomed attempt at self-medication,
But hold on—relief is nearly at hand.

Born without the enzyme that digests milk,
She supplies it via Lactaid tablets.
Born without the enzyme that digests thoughts,
Paxil prolongs her mood-regulators.

Behavior modification can raise
Serotonin, too, but not as quickly
As medications, like Paroxetine,
But it still works, half-effectively.

The highest zone is absolute happiness,
'Though even the best can slip to well-being,
And, sometimes, down into the bearable zone;
Next come the anger, apathy, and death zones!

Once you drop into the anger zone, the
Analytical mind cuts out, giving way
To the primitive reactive mind, a
Moronic state in which even beige seems black.

The simple reactive mind thinks that, say,
A perceived bad tone equals insult equals
Hate equals great anger equals lash out
Equals big fight equals kill equals death.

The mind is quite weak in the fighting off
Of emotions, for they have a direct
Pathway into the mind's eye—inhibiting
And suppressing the logic of the brain.

Emotions usually take sole control,
All logic relegated to the sidelines—
Being ineffective against a mood;
It's a wonder what we have a brain for!

Reason and emotion are not coordinated,
Each having a separate pathway to the mind;
That, then, is all there is to tell about all the
Miseries and follies of human history.

Sometimes, the way that we feel depends but
Upon chemicals—neurotransmitters
Like Dopamine and Serotonin that
Fluctuate, so—how meaningful are we?

Let reactions sail on by—just observe them,
But don't act on them. This puts some distance
Between you and your conditioned response,
A space which grants a modicum of free will.

PHOTONIC WONDERS

I took a snapshot of a moving photon;
It filled a pixel and was not moving on.
The photo told me not how fast it had flown,
So I assumed this info couldn't be known.

The photon was ageless at the speed of light,
As women are always young and always right,
For time had stopped; I thought time was movement.
Photons never get old or need improvement.

With my new contact lenses, I now can see
One photon, unmeasured by man—most need three.
It traveled 13 billion years, from the deep,
But what lights my dark head when I dream in sleep?

How come photons don't pile up on the floor
Under my lamp when it shines all the more?
Lucky thing, for where would I put them all—
Doing light housekeeping into the hall?

Is it light that defines space, as EM?
Do I see the light? What is lit in REM?
Is light the answer to the TOE's dark quiz?
Then wherever it reaches, existence is.

THE BIG BANG

We are ready to mix the Witch's brew
Of that endless energy making me and you...

...Stand back, for when we tickle the tail,
The dragon may laugh, cry, or ach-choo.

UNIFICATION OR SEPARATION?

Electricity and magnetism each
Lead to the other, being transformational.
They facilitate action and motion
Through EM's push-pull of regularity.

The strong force binds the atomic nucleus,
Barely beating EM's repelling force.
The weak force counters strong's stability
Through decay that promotes changeability.

Electromagnetism and the weak force
Unify when the temperature gets hot,
As during the Big Bang, and they oppose
The strong force as duality's balance.

What about gravity? Where has it been?
It needs matter and motion to exist
And so it is the blended result of
All the forces, a secondary effect.

Dualities seem to assist nature:
Good/evil, on/off, hot-cold, man/woman,
Up/down, left-right, here-there, past-future, and
So, none can exist without the other.

There can be no more unification,
For what One could be versatile enough
To form both the electroweak force and the
Strong—as different as the north/south poles.

TO BE OR NOT

The option not to be is unavailable,
Since all is here versus not here at all.

That is the cause, the 'Why' of existence;
The 'How' is potential possibility.

REALITY FABRICATION

Take a pencil and feel a texture with it
You seem to feel it at the pencil's tip;
But, you have no sense organs way out there;
So—the brain fabricates reality!

All you see is the inside of your head,
A model. You don't believe it, you say;
Well, it's the same model seen in your dreams,
With your eyes closed and you in darkness, too!

All we see are the insides of our heads—
A wide-awake dream. What's really out there,
I suppose, are waves and fields, which our minds
Sense in representative ways, like red.

Not only is seeing inside the head,
But also the hard-soft-texture of touch,
The scents of molecule shapes, and the sounds
Of air waves, again, as in a night-dream.

Absolute Reality is scentless,
Colorless, and quite soundless; however,
Sense organs detect waves and vibrations—
Such—all reality's fabricated.

A virtual reality can be
Enjoyed and directed in lucid dreams,
Where one can do anything at all, without
Injury or penalty, with real feel!

A simple four-way lookup senses taste
By degrees of bitter, salt, sweet, and sour,
And, likewise done, the three-way colors,
And ten-plus-way facial recognition.

Brains have parallel processors for form,
Texture, color, and depth—and a quick one
For motion detection—which all combine
Later as what we see in unity.

Consciousness is referred back in time a bit,
Like the tape-delay of a live TV show,
To hide the brain's processing time from us,
Making things seem to happen instantly.

The now is ten-forty-fourths of a second long,
The frequency at which events appear over
The horizon of consciousness, the succession
Of which gives us the illusion of time passing.

A hormoned hunger pang, midway between a
Pain and a withdrawal symptom, makes us
Run, run, run to eat—well, that's OK, until
Another hormone signals satiation.

Bonding hormones bring us closer together,
Pheromoaning into lust, love, and relationships
Spurred onward by love-made endorphins—and, so,
Yes, there can be chemistry between people.

People are but machines going the way
Of their brains, genetics, and chemicals
But, you, learning these secrets, rise above,
For, at least, you know you're a machine.

REALITY

Mind ever matters;
Matter never minds?

If all is minding,
Then nothing matters,
But, if matter does,
We truly live as real
What is and was.

FREE WILL

Do you control your thoughts or do they control you?
Could you, silly as it seems,
Just be falling, hook and line, for your thoughts?
Think about it—thoughts may tell you the answer!

The brain's decisions are determined by
Memories, associations, and
Learned behaviors right up to the instant;
So—our decisions are predetermined.

The "free" in free will has no real meaning,
Unless we take it to mean random, that
One's will depends on nothing but dice rolls;
What good would such a brain be anyway?

Can you start or stop your thoughts? In other words,
Can you will that which does the willing? Try it.
Oops, a surprise thought just came from the blue;
You did not will it—the will is unfree!

A mind is perhaps many little minds,
Each a simpleton awaiting control,
Such as when we eat, socialize, or fight,
None of them very complex at all.

The brain, with its hundred billion nerve cells,
Does all of our decision-analysis,
Only making its results known, at the last,
To the brain's highest level: consciousness.

People act, robot-like, since they know not
The why of what they do, for decisions
Are made blind, by brain networks, just before
They're presented to us in consciousness.

Consciousness comes three hundred milliseconds
After the brain does its analysis,
And, thus, has but last-second veto power,
If any, over what the brain comes up with.

Decisions are not made by consciousness,
Although, this fine picture in the mind's 'I',
Merely the brain's perception of itself,
Is fed back whole for future shortcutting.

Not much of what the brain does reaches
Consciousness, and even when it does,
The mind's last to know, being like a tourist—
For decisions precede their awareness.

First-level people have beliefs and desires,
But second-level people can have beliefs
And desires about their beliefs and desires,
Becoming able spectators of themselves.

Although our decisions of the instant are
Fully determined, and are therefore not free,
We may happen to learn something new—and make
Choices tomorrow we wouldn't make today.

Thoughts good and bad come and go, as the brain
Looks at itself without assigning values.
Still, lucky that others can't read our minds,
'Though forbidden thoughts are normal and sane.

If you try hard not to think of something,
Then you will just think of it all the more
So, if told to avoid impure thoughts, you'll
Think of people naked beneath their clothes!

We may fall for our thoughts, hook, line, and sinker:
Conditioned responses, reflexes, or
Overwhelming emotions, spurious,
Or ancient, planted by evolution.

When extreme thoughts arrive, uninvited, as
Most thoughts do, we veto them, saying "don't",
For while we can never will that which does
The unconscious willing, we have some "free won't".

We're all robots, but no one notices
Since there are so many different kinds,
Which, though making life quite interesting,
Obscures the fact that the will is unfree.

THE STARRING ROLES

Protons formed a massive star,
Via gravity,
And for quite a while
It fused hydrogen into helium,
Living a long and healthy life,
But its death would be
Even more spectacular.

In its death throes,
This massive star
Goes out with a bang—
First collapsing,
Then triggering
A supernova explosion
Bright enough
To drown out the light
Of an entire galaxy.

A shock wave of precious stellar debris
Hurtles outward into space
At tens of millions of miles per hour,
Containing the heavier elements
That will make up planets,
Form more stars, and even create life.

POPS OF POTENTIAL

I looked about, again, at the living lumps of clay.
Wondering where the Potter was at the end of the day?
Was He, too, back then, formed as a pot like me,
By that Grandaddy of All Eternity—Possibility?

THE AGE OF RUST

Aliens visit the Earth, thinking lush,
Only to find it covered with rust.

Their visit does not occur in future times
When the Earth is full of dusty rime,
But was way back in the youthful years
Before there was any atmosphere.

Bacteria have just begun making oxygen,
Discarding it as a mere waste product—
An unwanted poison to be jettisoned.

Everything on Earth that is capable
Of being oxidized becomes oxidized;
We've seen evidence of this rust
In bands of red oxide deep in the crust.

Only when this rusting comes to an end
Do the oxygen levels in the atmosphere rise;
They are only 1 percent after 2.4 billions years;
Now being 20 percent after double those years.

The aliens soon leave our planet,
That rusting junk pile that
Could never amount to squat.

They did take a photo
Of the Eiffel Tower, however.

THE VIRGINS CLUSTER

The galaxies of Virgo are strewn about...
The brand new ladies ever within without,
Like those glorious new flowers of May—
'Tis where Islamics go when martyred away.

AWARENESS

How is conscious reality real-ized?
What directs the spotlight of attention?
Who's the silent witness that can do none other?
Who is the knower that does the knowing?

'I' equals 'awareness' of the mind's state,
But who or what, then, is this 'I', observing,
Which is all 'I' can do—that transcends cells,
But needs a brain, being nothing in itself?

Awareness can never be an object
Of observation because awareness
Is the very means whereby we observe.
We can't see awareness since we are it!

We can seldom really understand awareness,
A subject, because it's not itself an object
That we can be aware of—for the only tool
We have to use on it is awareness itself!

Since the 'I' of awareness, that can but
Observe the mind's contents, is not a self,
It's a universal subject, a "wave"
Of unperceptive immortality.

In consciousness, there's no distance between
The thing observed and what is observing;
They are, in fact, one and the same, and so
It is that we seem to have a self.

'I' am not this body—or even this thought,
For 'I' am a part of space-time itself,
Although 'I' require a mind/brain to look—
For this, indeed, produces what 'I' look at.

The Soul of our Awareness is a basic
Property of space-time that can but observe
The mind (the contents of the brain), that is,
The portion that's currently on display.

Energy and Awareness are of the same
Ground of Ultimate Reality,
On which both quantum-jumps
And mind-thoughts are built,
Leading to the matter of brain neurons.

Awareness is all there is; look no further,
Its swirling energy gives a presence,
And a knowing, to waves and particles,
That leads, in total, to our consciousness.
(not really)

BEING NOTHINGNESS

Our parentheses in eternity
Flashes as a twinkling, but's extended
By time into a phantasmic life dream
That's existent the same as if it were.

A life dream's like a rainbow, not really there,
A false phenomenon become tangible
Through its being, the true true of the faux true,
Molding a genuine significance.

Life's indeterminate or not, the same
Being brought by the virtual as the true,
The mechanics being as incidental
As why 'color' chose its wave frequencies.

Life's here, like a virtual particle
Born this side of an event horizon
Of a Black Hole, realized by its presence
In the realm of what's been radiated.

There is no difference of what makes none;
Realism is now playing, the living film:
A reality show in the theater
Of the mind's eye, with the 'I' observing.

CONSCIOUSNESS

Why should the wetness of water result
From the mix of hydrogen, oxygen?
How can cells, blood, heart, and nerves make life?
It is just so. So does matter make mind.

Change the brain and consciousness changes, too.
Take drugs and the emotions change, as well.
Damage the brain and the mind's damaged, too.
Consciousness emerges only from the brain.

The brain is the mind, and vice-versa,
So there is no need for the mind to turn
The brain's water into wine, for there's
No wine that's separate from the water.

Consciousness is emergent from the brain,
A most fundamental phenomenon
Could it be the brain perceiving itself,
Something we might like to call the mind's 'I'?

Consciousness is kind of a fundamental force,
Like mass, space, and time, and, therefore, requires
No explanation—it just arises:
Mind: it matters; matter: ever mind!

Pain's not the same as the nerves that cause it,
Yet, mind, apart, couldn't conserve energy.
It seems that info exists in two ways:
Consciously and neurologically.

Nature's made of occasions of experience
Instantiated into consciousness,
Even for electrons and lower life forms,
'Though worms sense but a smudge of reality.

In identifying consciousness,
We often confuse what is floating in
The stream of consciousness with the water itself;
Such, we note not the sea in which we see.

The Midas-magic of our consciousness,
That quantum alchemist of potential,
Creates the Real from the Possible, for
Everything it touches turns to matter!

Consciousness is irreducible in terms
Of basic entities, so, most likely,
The intrinsic properties underlying
Physical dispositions are experiential!

Subconscious trains of thought vie for attention,
Dueling choirs competing for first place
In the mind's 'I'—consciousness—to produce
Future, for this is the task of a thought.

Consciousness mediates thoughts versus outcomes
And is distributed all over the body
From the nerve spindles to the spine to the brain,
A way to actionize without moving.

THE GOOD OLD DAYS

'You' were once a lucky shrew, darting all about,
But then attached to a favorable evolutionary line...
Every single one of your forbears on both sides
Being attractive enough to locate a loving mate,
And, they, fortunately, had the good health to celebrate!

OUR ASTOUNDING FAMILY TREE

None of them were deflected from delivering
Their tiny charge of precious genetic material
By being devoured, drowned, starved, stranded,
Stuck fast, fatally wounded, or left empty handed.

QUANTUM PHYSICS

In the eerie quantum world, all possible
And potential realities
Exist all at once, in superposition,
Until one emerges into reality.

Electrons as waves are all spread out, and,
As such, are nowhere, having no position,
But, they have direction. As particles, they
Reside someplace, but have no momentum.

Without a position and momentum
At the same time, electrons have no
Objective reality at all.
They go from here to there with no in-between!

Sub-atomics exist everywhere,
Yet, nowhere, until they are measured, and
Seen by observation, so, until then,
They're as the selfless states of meditation.

Photons seem to know they're being measured,
Since they send out an offer wave through
Both slits at once, but receive the handshake
Of acceptance through just one slit, then go there!

One photon, unmeasured by man, does go
Through both slits, showing wave interference
Because, well, it splits the world-path in two:
One that we stay in and one that we don't.

Minds seem to sense in another dimension,
Collapsing possibilities into reality,
A quantum mind tries out new ideas
Through scenarios of consequences.

The back-action of a particle on
It's pilot wave, in concert with the wave's
Guidance, creates the élan vital,
The stuff that consciousness is made of.

A particle's action is often zero
On its guide-wave, be-coming randomness, but,
At the mind/brain level the brain feeds back
Into the mind's guide-wave, and vice-versa.

The quantum substrate's the mother of all,
Of both matter and consciousness,
Using like subatomic nonlocal effects
That yet link particles which are far apart.

UNIVERSAL INSURANCE POLICY

The underwritten Underwriter
Of this universal wave of matter
Covers all loss and liability,
Guaranteeing payment,
By dipping into Possibility,
Issuing both the credit and the debit.

FANCIFUL DUO

Our never-the-less real reality is brought forth
By our consciousness, which, paradoxically,
Is real as well. It's just that reality, although definite,
Is malleable in that it is part of a feedback loop.

For example,
"We" created the beginnings of the universe,
Which in turn evolved into us.
There is an actual realm of instantiation
As well as a possible realm of information,
Each of which can formulate the other.

This gave direction of a kind to what otherwise
Would have been too fortuitous to be here.

MEDITATION

People can't usually ever see
Further than an order of magnitude
Beyond where they are rutted, but...
Some can intuit ultimate reality.

It said, in my dreams, *Of ever waking,*
It's hard to convince you with dream-language,
As when, in wakeful reality,
To tell you of that which is beyond telling.

During meditation, one clears the mind,
And so, then, there's no real self, just one quale
A near nothing that has little need to be;
Is this what-it's-like to be a pure soul?

Physics, once more direct, is now but an
Immaterial science of math-shadows,
While, mysticism, once but a fogged notion,
Now's the direct observation of the Light.

Meditation shifts intention away
From controlling and acquiring
Toward acceptance and observation:
You take-in instead of acting upon.

Enlightenment's not grasped or possessed
Acquisitive aim locks the secret out—
The form of consciousness that one starts with;
This is why "the secret protects itself".

The Spiritual refers to profound connection,
Though not through visions or ecstatic emotion,
But with the experience of connectedness that
Underlies reality, and nothing more.

Meditation relieves the survival self,
Shifting attention from acting to allowing,
From emotional identification to observation,
From instrumental thinking to receptive experience.

Meditation, renunciation, and service
Are not really mysterious, just different
From the usual object-oriented approach...
Mysticism is modern and ancient, not esoteric.

In serving the task, one forgets the self,
And accesses life's connected aspects
That go beyond one's self-centered consciousness,
The survival of mankind being at sake.

Awareness is the ultimate being,
Fundamentally connected with 'Soul',
And cannot be known in terms of worldly
Objects—it's like, well... you have to be there!

The connectedness of everything to everything,
A rudimentary perception in and of itself,
Experiential in its ultimate physical disposition,
Facilitates our consciousness of exterior through interior.

Or, possibly, probably, the quieting
Of the brain's self-boundary ID center
Via focus on mantras, hymns, or prayers
Is but a neurological effect, nothing more.

THE END

Oh, Olongapo, fleshpot of fertile flora,
Pinatubo has resealed your box pandora.
Fiery ash has frozen your beauty in time—
A poem in stone, like Sodom and Gomorrah.
...
Obliterated by a war nuclear,
The Earth exploded in blazes solar!
Said a child in a galaxy afar,
"Oh, look! Look at the pretty shooting star!"

WORLDLY LOVE

I am thy moon, thy constant satellite,
Thy crystal paramour of day and night.
Above and below, and within thy sight,
I whirl around you in loving delight.

In a magnetic dance I whirl and twirl,
Attracted to you, the liveliest world;
Around you as a necklace I'm aswirl—
Wear me as thy crystalline gem impearled.

Wherever thou orbits 'round Apollo,
I must twirl and whirl, hurry and follow;
Dust I gather, meteors I swallow,
Ranging far and wide through space not hollow.

Thy romantic beam, like Cupid's arrow,
Pierces my heart and kills my sorrow,
Injecting life and love, for tomorrow;
Henceforth, I'll shine with this light I borrow.

Around you I whirl, a necklace of pearl,
Trailing afterimages of my world,
Adorning you, thy bosom bountiful,
With crystalline gems from another world.

As twin planets, our orbits must convolve;
Into each our tidal motions dissolve.
Around a common center we revolve—
The focus from which our passions evolve.

As twin planets, each other's way we pave,
With the push-pulse of the graviton wave.
We're captured, but not as each other's slave,
For to the sun our orbits are concave.

To your lines of flux my path I align—
I'm your constant paramour, crystalline;
Your world pours life on mine, on mine!
Dearest Earth, I must be thine, must be thine!

A magnetic beam emanates from thee,
Attracting me, holding me, kissing me;
Tidal love washes freely over me,
Linking you and me for eternity.

Basking warmly in your reflected light,
I'm bright, oh so radiant in your sight!
In the love and light of your spirit bright,
I need not ever face the endless night.

Your vibrations travel without a sound,
Circling from all directions to surround;
This affection touches me, 'round and 'round,
And closely binds me to you—I'm love-bound!

We're as different as midnight and noon,
Yet drawn close by the force of Earth and moon;
As lovers, we merge in a sweet eclipse,
When world meets world as a kiss on our lips.

Oh, as your shadow of love covers me,
I am full, so full in the shade of thee;
When we overlap, that union is us
The you is in me, the me is in thee!

As moon and Earth we bathe in radiance,
Cleansing our hearts in love's grand alliance;
'Round and 'round each other we dance, entranced,
Revolving in the whirl of our dalliance.

MAGIC?

How could a random event manifest itself
With no conception of it somewhere else?
Wouldn't this be something become of nothing?

HIGHER CONSCIOUSNESS

The three lower consciousnesses that are
Obsessed with the securing of objects,
With the chasing of sensations, and with
Power/control will never ever be enough.

There are NO actions of people that can
Justify our becoming irritable
Angry, fearful, jealous or anxious if
We give them our unconditional love.

If we don't accept the unacceptable,
Then we lower our level of consciousness
Our response will mirror their uptightness—
Which can spread the bad moods onto others.

Conscious Awareness, which can but witness,
Is a safe haven from which to observe
The drama of our lives playing in our minds,
Granting us a sobering distance from it.

From a safe subjective place that's free of fear,
Our soul, our conscious awareness, can witness
The strange thoughts and emotions that surface
On the mind, sent by the subconscious brain.

Putting ourselves in the place of others
When hurtful things are done to us,
Expands our consciousness, compassion, and love
Since we can come to know why they did it.

When we converse with ourselves, it is our
Higher Consciousness—our Conscious Awareness
Or I, that questions our lower consciousness
Impulses toward securing, sensation, and power.

Seeing the big picture of life and its stages
And connections lets one not get annoyed, say,
At being cut off in traffic, for s/he
May be old, learning, lost, growing, or angry.

Putting the needs of others ahead of
Our own produces the byproduct of
Happiness and reduces stress, for we
No longer have unrealistic expectations.

There's NO LIFE in the dead past, just history,
Nor in the imagined future, a mystery,
But in the here and now, life just arrives
Its a gift—that's why its called the present.

THE BALANCE

To infinity existence cannot be,
As all is not of great density,
But of some amount
Of something that forms our account.

ON THE EDGE

Existence extends its electromagnetic preach
As far as its atomic influences can reach;
Beyond all of that there is Nothing there
But the naught of very thin 'air'.

Nothing has no real existence
In some places, and far 'beyond' here.

TWO WAY STREET?

Life need not be either real or just a rental,
For consciousness could be funda-mental,
As that like matter, too, being is and was,
Arising whenever our brains' information does.

WE ARE THE TOE—
All the fun and mental forces unified as one.

TRUE COLORS

We are the Eternal Smile of Being,
The Joy of the Universe's Creation!
In us the Cosmos has come alive
And has evolved into our consciousness
From primordial matter and energy

We have arrived! We are the Cosmos itself.
We are the Universe—life from Stardust!

We live but for one of Eternity's heartbeats,
Borrowing Life from Death for just a while.
All that we are we owe to Time, Death, and Stars.
Truly, from the Stars cometh our help.

The Stars are the creators of atomic matter.
Within a Star's heart, matter transforms itself
And gives off energy—this is why the Stars shine!

Death is the ultimate evaluator
And the director of all evolutionary progress.
Death selects the wise from the silly;
Death chooses the useful from the useless,
But, it takes Time.

It is this long yardstick that sticks in our throat.
For what seemed like Forever,
Our sleepless spirits have waited to catch light,
Life, and delight from Heaven's smile.
Finally, we are so lucky and we live.

We stand atop the pinnacle of Nature's tireless toil
Which has at last brought forth our souls
From that black and endless eternal deep.

What a joy to Be!

In what far and fiery depths of space
Burnt the fire of your Spirit?
In what distant Stars was born the gleam in your eye?

Know it well, for one day Death will ask you
"What did you do all of your life?"

But, for now we are alive.
Our mind and senses interpret and disperse
The base Reality into the colors and sensations
Of the phenomenal world.

We can become either rainbows or ugly stains!
Our minds, like Shelley's prisms of many-colored glass,
Strain this white Radiance of Eternity
Into our life—until Death tramples us—
And back we go to stardust
After relentless time has wasted us away.

Yes, our creators of Time, Death, and Stardust
Must also write our epitaph;
They devour us in order to return
That life-dream which was lent to us.

But, here we are now, and perhaps we come to know
That the simpler things in life are still the best:
A glass of water from the well in the morning;
To love, laugh, and sing with family and friends.

And so we live out our lives with honor and love,
Kindness and generosity—these are our true colors.
Life for the sake of life! Good for good's sake!
Enjoying everyone and everything and every season.

Many think that they are more important
Than they really are, that they deserve some reward
Of a divine destiny in Heaven where their every whim,
Wish, and fancy can be fulfilled for all of time.

Well, to me, such endless satisfaction and pleasure
Sounds really rather prideful, wishful, even decadent.
The ultimate humility is for us to realize
That we are no more than electrochemical organisms.

Are we quite lucky and fancy organisms? Oh, yes.
Are we specially created by a Master? Oh, no.
We are the embodiment of the Cosmos
And are ever the results of natural laws.

Death will be forever, but man,
With his exaggerated view of self-importance,
And, not wishing to see a final end
To his glorious life—and I can hardly blame him—
Desperately grasps for immortality's promise.

For me, I will continue to catch life's joy and smile
And will bathe in the light of its constant sunrise.

On my last night on this Earth
I will not be haunted by regret
When the Sleep of Death comes
To take me to Corruption's dim dwelling place—
For I will know that I lived for color and smile.

And what of the Stars?
They remain, as Eternity's Love-lamps,
Representing our good works and deeds,
Which even the fathomless night cannot quench.

Perhaps one day, at the end of forever,
The Stars too will die and grow cold when
Time conquers all; but, as long as they live
They will shine and radiate the hues
That paint the colors of our ashes
Reborn again on the phoenix wings of Time.

NO FAULT

Since we can't know whither whence we came,
There would then be no paying for blame or shame,
So, again, life is a freedom to be—a gift, the same
As as if we believe that from Loving we came.

WHAT REALLY HAPPENS AFTER WE DIE?

We die 'little deaths' all the time.
Our atoms change,
Some of our memories go away
And some new ones reappear,
Although I realize that it
Is the core of memories
That defines us as us.

It's just that we are hardly
The same person now
As when we were much younger.

We had 'death' before birth, too,
And now there is life after birth.

Is there life only during life?

If one had amnesia and began
Learning the world anew,
Then one might say that
One as the previous person
Was 'dead' and that it is
Our new life that counts,
One not even missing the old one.

And, while the 'big death' is much more
Than any of these 'little deaths',
It is that our atoms then go on
To reside in a new person eventually.

It's not like there is any continuity of memory,
But more like that any narrative will do.

As for really knowing all, that is, the TOE,
Meaning that we know that we can't know it all.
This is a relief and so then we can go about our life,
With the ultimate freedom to be.

FUNDAMENTAL POSSIBILITY

Time, space, stuff, change, and form were real-ized from
The Fundamental Possibility,
Becoming our penultimate reality—
One possible from all probabilities.

Our reality came not from nothing,
But existed always as possibility,
One that amounted to something workable,
Among all in superposition.

No form of our penultimate realness
Could have existed alone before
Everything's options were known-all-at-once,
For what could have made the choice among many?

Nor came it from an absolute nothing,
Since there can be no such "thing" at all,
So, since either way is impossible,
Fundamental Possibility <u>is</u>.

This ultimate basis of reality
Though not much like our local reality,
Is hinted at by quantum physics—
It forms reality real as can be!

So how else could it be, for particles
Do appear and disappear from somewhere,
Going from here to there with no between,
Manifesting from no-where to now-here.

I'll follow every single avenue,
Whether it's brightly lit or a dark alley,
Exploring one-ways, no-ways, and dead-ends
Until I find where the truth is hiding.

Since we all became of this universe,
Should we not ask who we are, whence we came?
Insight clefts night's skirt with its radiance—
The Theory of Everything shines through!

Some simple substances gave rise to everything,
Chosen as probable above the rest—
Known all-at-once that it would be the best—
The most promising—the possible ones.

As to how complex, there is no limit
But to collapse into a black hole;
The smallest of all is the planck distance,
So size is absolute, not relative.

Like the moon, challenge night and gain the light;
Like the rose, suffer the thorn—gain the fragrance;
Of life, surrender to live forever—
Enlightened more than a thousand suns.

World does not pass by—you pass through it;
Clear your being so the treasure may arrive;
This spirit sparkles of a different light—
The gemstones are of a different mine.

THE PROCESSION OF
THE CONSTITUENTS OF REALITY

Sad Yesteryear, Forever, and Everywhere,
They all came, to weep for Nobody Nowhere,
With Why and How, Then, Now, When, & What & Where,
Led but by their tears and sorrow. Your posts zing
With things that "none" can bring: Everything.

FEELING IS BELIEVING?

I have felt the spirit of Santa Claus
And yet he is not the real cause
Of the presents delivered
And made by elves at the north pole.

NOW AND ZEN

Everything that is part of us—
Our cells, tissues, organs and organ systems—
Has come about over billions of years
Because it proved successful
In the great survival stakes
During our perilous evolutionary
Descent (ascent) with modification.

The brain, being no exception,
Evolved, in part,
To allow a creature to learn
From what happens in its life,
To retain key elements that
Could influence future actions.

We are geared for self-preservation.
We will do anything to avoid facing the possibility
That who we are now cannot continue.

We ourselves are mainly the cause
That we are interested in.
The self is preoccupied with staying alive,
Which is why our species is still around today.

It is a prime biological function to be afraid of death,
And, so, the self, as thus contrived,
Is able to fully play its crucial survival role.

We want to equip our brain with a soul
That offers us an escape when the brain dies
Since the self cannot come to terms
With its own extinction.

From a subjective standpoint,
We are all born equal and undifferentiated
(Before that, 'we' were dead),
But, as mature selves we make a distinction
Between the individual and the surroundings.

Still, the brain keeps changing throughout life,
In a pattern of the shifting flux of its neurons;
We gain and lose memories and feelings,
Essentially creating a new person over and over again.

The self is thus not so rock solid as it seems.
These moment-to-moment changes differ from death
Only in degree. In essence, they are identical,
Although at the opposite ends of the spectrum.

So, we are not static things.
Other neural networks will come to be in other,
Future people, albeit with an "amnesia"
Of what went on before in
The brains of the previous others.

Why should we be happy about this?

We never can be, because the 'I' cannot operate
Outside of its own boundaries.
The only viable alternative is to think of a way
In which it is possible to ever continue on.

What will it be like to be a part
Of someone else after we die,
With our own particular
Narrative of life cast aside?

This is the 'zen' of now and then and when.

ROOTS

Evil roots from the making up of 'good'.
Farewell, false preachers of all varieties
Of methods and dogmas that say "do this or else".
Your "good's" are flawed by being made up.

THE SEED OF LIFE?

Mikey was a unicellular microorganism,
A microbe, one of the bacteria
That were called 'extremophiles',
For they were capable of living
In extreme environmental conditions
Of temperature, pH, salinity, pressure, dryness,
Radiation, and even with no sunlight or oxygen.

They even loved chomping on plutonium,
The deadliest substance ever known.

Mikey's ancestry went back 4 billion years,
He being among the sturdiest creatures on Earth,
Those that had also become its master,
For humans couldn't live for
But a few minutes without bacteria.

Mikey thought that he might go to Enceladus
For a balmy vacation where life was easy,
Always with a pool and party not too far away.

Enceladus is a small satellite of Saturn
And is a geologically active moon world
With some wondrous scenery
Of spouting volcanic plumes,
Even having a bath of water within and below.

Just about then,
For sometimes wishes do come true,
A huge meteor impact struck the Earth
And thrust some material into space,
Including Mikey and friends,
Who then resided rather dormant
In a rock that protected them
By acting as a shield
Against solar radiation and cosmic rays,
Not that this would have bothered them a whole lot.

Eons later Mikey and friends
And their rock of a spaceship landed on Enceladus.

Mikey stepped out of the rock
And onto a tiger-striped surface
Where the temperature was about -359 degrees F.
A tiny shiver almost began to undulate through him,
But, he shrugged it off.

He was hungry, though,
Not having eaten for millions of years,
Except for a few bites of iron—
And so he was really only running quickly
At about half-speed.

His friends followed excitedly,
Covering over 100 kilometers in a few minutes.
They paused every so often to gobble up some dust.

They were taken aback for a millisecond
When they spotted a fast food restaurant
With a sign that said
'Billions and billions of bacteria served here'.

"Hey, there is native life here,
Just as we'd hoped" said Mikey.

"What a tropical paradise!
Hey, there are some hot springs.
Let's take a dip with the sexy native girls
And then kick back and relax."

They frolicked and swam all around
For a few thousand years...
Until a very large eruption
Sent them all far into space.

After a billion years or so,
They landed on the 4th planet from a sun
In a solar system far away,
Seeding it with life that became human-like
Within a few more billion years,
Although their were some differences in anatomy.

— *Mikey*

THE FISH WHO ALMOST EVOLVED MORE

On the road to Kingston, NY, the other day
I was happy to see that the rains had returned
And that the drought was ending.

I stopped to do a little fishing along the way,
And it reminded me of a fish
I'd caught during the dry season.

I'll tell you about it now.
It took me awhile to get over it.

It was so dry that I could walk
Across the reservoir and the creeks.
The water was shallow due to the drought,
And most of the fish were swimming sideways
So they could stay under water.

Then I saw a rather amazing sight:
One fish was leaping from puddle to puddle,
Sometimes crawling across the dry land in between.

I threw away my fishing pole
And caught this fish with my bare hands,
Thinking of the delicious fish fry dinner
that I would have that night.

I put the fish in a bucket of water
In the front seat of my car
To keep it fresh during the long drive home.

Every so often that fish would poke its head
Out of the water bucket and look at me,
Sometimes even trying to jump out.

Finally, it did get out of the bucket
And sat on the seat next to me.
It was then that I realized that
I could never eat this fish.

About the same time, a brilliant idea stuck me:
I would train this fish to live out of water,
And make a pet out of it,
As the fish seemed to already
Have inclinations in that direction.

At home I put the fish in a barrel of water,
And sure enough, it tried to jump out.
So, each morning I would take it out
And put it on the grass,
Which was still wet from the dew.

Then, when I could see that it had had enough,
I would put it back in the barrel.
Each day the fish seemed to last longer and longer
Outside the water barrel before getting listless.

After a few months of this,
The fish didn't need much water at all;
As I walked along the road in the morning,
It would wriggle along beside me
In the wet grass in the shade.

Later, when the day became really hot,
I would give the fish a drink from my water jug.

After a few more months of training,
The fish was able to flop and sort of 'swim' along
Down the middle of dusty roads.
And when I offered it a drink, it refused!

We even went to the beach together;
Of course; only I went swimming—
The fish just laid on the sand, getting a tan
And enjoying the breeze.

One day it was over 120 degrees
And the fish just had to have a drink,
So I gave it a dry beer.

Other than that,
The fish never touched water anymore,
Having become a land animal.

What a lovable pet!
It slept with me, saw movies with me,
Went out to parties with me,
Chased down tennis balls
And brought them back to me,
Rode on the back of my bike, etc.,
We were inseparable!

But then a tragedy happened:
We were walking down the road together one day,
And passing over an old bridge;
Suddenly my fish fell between some loose boards
And on down into the creek below and drowned.

THE DRIFT

I drifted down the River of Eternity
Born and borne upon its meandering current,
Floating, bobbing, and wafting unto this life,
Now here wandering, puttering, and dawdling,
Often digressing, deviating, diverging,
Veering, and getting sidetracked.

All eventually piled up, banked up, heaping up,
Accumulating and gathering the bulk amassed,
Yet ever shifting, flowing, and blowing,
And relocating, until I caught the drift...

Of that gist, essence, and meaning,
The drift of the shift,
From the sense, the substance,
And its significance as the mound that
Accumulated into me—that thrust, tenor,
And implication imported into me,
Which spurred the intention
To direct the course of history.

THE NIGHTMARE

She, looking like Melanie of ToeQuest.com,
And still in her pajamas,
Grabbed her purified water bottle
And hopped on the bus.

Glad to see her,
I waved her over to my seat,
For she was my guru.

I was also in my pajamas,
For this made the yoga
Of our meditation therapy easier.

"This is not a real bus,
Nor is is really moving," I offered.

*"True," she replied, "we are dream characters
In the dream of the Perfect Awareness."*

"It just all plays out in our consciousness,
Kind of like a movie."

*"Yes, nothing comes through the senses
Or from the brain or any thing like that;
It's life's soap opera channel
And there is no remote control."*

"It goes as it has to go; all is illusion,
But we aren't fooled at all."

"No, we are foolproof."

"Why are you wiggling all around?"

*"I have to pee; it's a dream pee,
But it is still a dream hurt."*

"Well, you could get off at the next stop
And go into a building.

I'll see you at the class."

"OK."

"I'll tell you a short joke
Before the stop comes:
If you are Russian before
You get to the bathroom
[water closet to you]
And you are English
After you come out,
Then what are you
When you are in the bathroom?"

"I give up; the dream didn't tell me the answer."

"European!"

"Ha, ha. Now I can't wait for the next stop!"

"Just ask the driver to stop near a building
Or a gas station and let you off."

"OK, I'll see you in a while."

She had to pee so bad that
She ran straight for a building
And rushed in right to the bathroom
Without anyone even noticing;
Nor did she notice what
The name of the place was.

When she came out of the bathroom
She was English again.
An orderly stopped her, restraining her.

"I'm sorry. I had to go."

"You need permission for that.
Now let's get you back to your room."

"What? I don't have a room. Where am I?"

"Why, of course, you are a resident
Of Chesterfield Mental Institution."

"No I'm not. I just got off a bus."

"We hear those kinds of stories all the time.
Where's you room?"

"I am sane," she said with a dry mouth.

"Would you like a drink?"

"No, I only drink a special kind of water."

"Oh, a special kind?
Then maybe it is in your room."

"I left it on the bus."

"There's no bus stop here.
Let's get you out of the lobby."

"I don't belong in this place."

"Then why are you wearing pajamas?"

"For meditation therapy."

"Therapy? Well, I can get you to that."

*"You don't understand. I am normal.
My God, what a turn this dream is taking!"*

"A dream?"

"Yes, nothing is real; all is a dream."

THE NO-BEGINNING OF FOREVER

"Hey, taxi," Austino waved.
It pulled over and he was saved.

"My 'no car' isn't working."

"What's the problem, parking?"

"It doesn't work because, wear,
I don't have one; it's not there."

"Where to, maternity?"

"From here to paternity;
I wish to see eternity."

"Which shall it be, how?
The eternity up to now
That had no beginning bow,
It stretching back forever,
Or the one yet to come, never,
That will ever go on toward infinity?"

"The road to antiquity."

"Good choice, for what we are now—our yore,
Is made of all that which came before,
Plus, the future of the universe
Isn't what it used to be."

"OK, Yogi. Can we get to either one?"

"No, for no matter how time we pour,
Either fore or aft, to beyond or before,
We'd still have an infinite way to explore."

"Know any good short cuts?"

"Yes, the dark avenue of Possibility;
It's back at the always beginning."

"There was a beginning?"

"Yes, but only of the universe."
But not of its cause, itself, we reach,
For no causes can be further beneath...."

"...Or causes would be ever regressed."

"There can't be real stuff around
That is already defined, with no ground,
With the properties of its nature
Such as its size, shape, durability,
Amount, capabilities and so forth,
Without ever having had any definition."

"What if the real stuff was eternal,
And not of anything maternally paternal,
Being unbreakable—
And therefore unmakable?"

"Still, what defined its nature
And its amount and all that's sure"?

"Yeah."

"So, our actual cannot be defined
Without ever having been defined."

"That's a relief; a basis, a 'plan';
Otherwise, all never began.

Things could never be made
Before they were ever made.
How long will it take to get there?"

"Well, relatively speaking,
It's only a small parenthesis
Of an 'eternity' away, a trifle,
*But a trillion**trillion years ago."*

"We need a better shortcut."

"We'll travel through the wormhole
That is at the center of the galaxy!"

"Sounds dangerous, for there's said
To be a black hole there, glowing red."

"Yes, there is."

"Some wonder if the egg
Of this black hole was pegged
Before the chicken of the galaxy
Or vice-versa."

"Well, on a previous row
Out into space, long ago,
I noted that in the past times
That the galaxies were 30 times
More massive than their black holes,
While present-day galaxies whole
Are 1,000 times heavier told."

"So, those holes in the past
Were still growing, to last;
Thus, the black hole came first."

"Yes, although I can't imagine, wacky,
How a black hole could build a galaxy,
For one would think that it would tear it apart."

"Yet another mystery tolled.
So, what is in a black hole?"

"Nothing, or rather a lack of space."

"Oh, wow, how does that work?"

"Since the speed of light,
Which is normally 'c',
Is zero in a black hole,
There is nothing happening there
At all and so there is nothing there."

"So, it is a hole in the fabric of space?"

"Yes, so then we can zip right through it."

"No thanks. I don't trust black holes.
They gobble up whole stars like coal.
How else can we get to the beginity,
This so-called land of possibility?"

"We can travel faster, though it's ever nearby,
For there are no speed limits around that 'place';
In fact, there are no laws of any kind there,
For what could have established all,
The laws before creation of the laws?"

"Indeed, it's a place where anything goes."

"Yes. Or you could think of it
As where every possible law exists,
As well as not any that exist."

"Yes, Everything sums to Nothing."

"True. And neither,
By the same reasoning,
Does it have any form."

"It would have to be formless
Since what it does is make form."

"Now you're getting it."

"No time there either—It's all at once,
For there are no forms to move about."

"Yes, this Possibility just is,
It never having been created."

"But it would need time even to create time!"

"Nope, all was/is done in an instant."

"So, Possibility is an all-at-once
Superpositional thing like that sum
We see that is the nature of the quantum?"

*"Yes, and once it makes the all,
Then there's time, form, and laws."*

"Is it a Great Mind?"

*"No, for a mind is an established
Complexity of lower parts that compose it."*

"How did Possibility think then?"

"It was like Yoda said: 'No think, just do."

"OK, Yoda, what did it do then?"

*"It did everything plain,
For it was unconstrained."*

"It took every possible path all at once?"

*"Yes, and our universe was one
Of those paths that worked out,
As it kept on evolving longer."*

"And the other paths?"

*"Perhaps some were flops
And perhaps some others made it,
But here we are, now stuck in traffic."*

"What? How come?"

*"They made a pedestrian mall
Of Broadway. How to get around it?"*

"But it's the only diagonal street in Manhattan."

"Yes, but it kind of keeps coming after us
As fast as we try to get around it."

"Can we take another shortcut?"

"We'll take the Lincoln Tunnel into New Jersey."

"The wormhole?"

"Now, is that any way to talk about New Jersey?"

"Yes."

"Yeah, ain't it the truth."

They skipped the Lincoln Tunnel
And found another funnel
Back to the beginning of forever.
It had this sign in front of it ever:

Neither Nothing nor infinity can be,
As Nothing is not there and so 'it' cannot be,
And infinity can never be reached,
Being a hopeless sequence never completed.

Something real had to be, but not forever so;
Thus, there was was creation of things that grow,
Via the 'something' of no laws, form, time or fee—
That which needs nothing before it: Possibility.

ENDLESS PLAY

I still jump rope in an eternity that has no ends,
But, if I realize that nothing is holding it up,
As like in cartoons when one notes nothing below,
I might trip or fall, never to return again.

THE END OF THE EARTH

The Asphodel sustains the Dis dwellers,
Where they rest beyond that fatal river—
There the wretched shades drink forgetfulness,
And to oblivion sink without distress.

Fireweed grows from Hell's sulfurous embers,
As does Purple Loosestrife—dead men's fingers;
But wildflower air revives the dead—and then
Those happy souls can thrive on Earth again.

Charon was withered, wan, and skeletal,
Although eternally grateful for his immortal life
And steady job of ferrying the dead across the river Styx
In their transition from life to death to forgetfulness.

As Earth was the only planet he'd come across
With such promising higher life forms,
Charon had grown rather fond of its inhabitants,
Even though he only saw but the worst of them;
But, even from this he could extrapolate
To the qualities of the best.

Charon did his job well, professionally,
Although it was ever so dreary
With the endless darkness of wasted lives
And the grim and gloomy skies all around,
For this land always had
That same gray and leaden feel.

He ferried on, though,
For his own life was precious to him.

The soon-to-be really really dead never said much,
For what was there to tell after an empty life
That had often turned to deep regret;

So, Charon did not prompt them for information,
For this was not the thing to do
At the time of their passing,

So he was always most
Courteous and kind to them,
Even to the most evil of the darkest,
Doing his task as well as he could.

It was not that Charon was afraid that
His undersized master of the underworld,
Pluto, might be watching,
But that he had the extreme clarity
To duly serve the task at hand,
A testament to his character.

Charon had been quite alarmed lately—
What with the numbers of the hellish-souls-to-be
Climbing into the millions in such a short time,
But, he had been through this kind of rush before,
With the doomed and damned of other planets
That had been consumed by their suns
Or had undergone other such catastrophes.

He just used larger boats
And patiently took his time,
For he had all of Eternity.

Of course,
Charon could and did feel deep sadness,
But he didn't show it outwardly,
Even when the numbers from Earth
Increased a thousand-fold again.

A few of the now billions
Of depressed Earthling souls
Had enough energy left to mumble a few words
And so he was able to glean from them
The latest happenings on Earth.

In 2012, the predicted exponential surge
Of melting ice from global warming
Had quickly inundated all of the coastal cities,
Many of them large centers
Of population and commerce.

Everyone who could possibly make it
Had to retreat inland,
Creating the largest mass exodus in history.

As the heat rose to unbearable levels,
Many had begun living in their basements
As the Earth's infrastructure
Began its eventual collapse.

Millions eventually headed north
Towards Canada and Siberia,
But had to retreat when the ice caps totally melted
And formed the great Ocean of the North;
Most did not make it.

No one but the ignored physicist mathematicians
Had predicted that the end
Could come into sight so quickly.

Then came the dreaded polar shift
That made global warming seem but a small note
Compared to this new and Darker Symphony.

The Earth was thrashed with storms
The likes of which it had never seen;
Electricity was completely out all over the world,
But for a few nuclear powered areas that didn't last.

No one could drive very far,
Even on their last tank of gas,
For the roads had melted,
Along with the tires of the vehicles,
And, if the vehicles stopped,
They'd find themselves mired
In the meltdown of the asphalt.

Food would no longer grow very well,
Even in once lush gardens,
In the amounts that were needed,
And, as the heat rose further,
Into the 140s, plant growth ceased altogether,
Although a new but rare

And expensive form of food pill
Extended life for some of the rich
For a short while.

Charon, had, of course,
Seen much of this kind of thing before
From the many other solar systems
And galaxies on which life had formed;
But Earthlings seemed to have
A special charm and hope
Above and beyond the other alien races;

So he rowed and ferried
And deposited them on the far shore,
His job and life forever continuing
In a place with no color,
No joy, and no future—
On the shore of the land
On the edge of oblivion.

Charon had depths of compassion,
But many passengers might
Many thought him stoic,
Although they were mostly
Beyond the capability.

A sign on the opposite shore said:

Abandon Hope All Ye Who Enter Here

Billions more arrived
In the gray land all too soon
And Charon learned that
Either madness or desperation on Earth
Had caused a nuclear winter all over the planet,
Bringing on a deep freeze that few could escape.

Perhaps they were trying
To combat the ultimate heat,
Which would have been
But a cool breeze in Hell.

The polar shift had greatly
Added to the deep freeze.
A few of Charon's still speaking
But chilled customers
Even expressed a longing
For the legendary warmth of Hades.

Charon, stalwart and reliable, rowed on steadily,
Ever steeling himself to the misery.

Finally the masses slowed and dwindled
To a few dribs and drabs over a few years
And then there was no one for several years.

A lone man appeared on the shore near the ferry dock
And Charon readily approached the man,
Something he had never done before.

They had a long and hearty talk,
For the man was animated
And not at all like any of the other wretched souls.

"How is it," inquired Charon,
"That you are full of life and seem to be a good man
But have been sent here?"

"I am not a bad person in any way," the man replied.
"Actually, I just spent some time in Heaven.
I found out there that my sweetheart
Was sent here to you,
For she was a suicide
And so was destined here;
However, I had promised
To be with her forever,
So I chose this place
Over Heaven out of my love for her."

"Extraordinary," exclaimed Charon.
"I knew the Earth had
A few good men and women—
I've not seen very many clues
Of that elsewhere in the universe.

Did you colonize space—
Will your species continue and flourish
After your Earth bids farewell?"

"I'm afraid not," replied the man,
For too many needless wars intervened
And this greatly delayed our space program."

"A shame," said Charon,
But is there any hope left on Earth,
I mean, are there any others still about?"

"I am the last," the man answered slowly.

The first tear of Charon's long life
Rolled down his cheek;
Nothing had ever made him cry before:
Nothing had ever made him weep.

(Rewritten from Lord Dunsany's brief sketch)

ARISTOTLE

"I abhor vacuuming."
—Austino

"Hi, Austino," said Aristotle,
"My universal rug is dirty and my vacuum sucks."

"Good. That's just what it is supposed to do."

"I got it at a dollar store.
I mean that it really doesn't suck at all."

"Oh, then keep it if it's so great."

"At least one vacuum is completely empty."

"Then of what use is it?"

THE FOREST OF ORIGINAL GROWTH

What would it be like to stumble across lands
That no one else had ever been to,
And how could you know that?

After reading Sir Conan Doyle's 'Lost World'
About dinosaurs on a sealed off plateau of a volcano,
I wondered if there were any more undiscovered places
That the paths least followed could lead me to.

So, while at the Earth Summit in Rio last month,
I forayed into the uncharted regions of Brazil,
Having chosen from a map the remotest area.

After various vaccinations and preparations,
I trucked my one-man helicopter
To the last way station,
Loaded the extra gas tanks onto it
And flew into the heart of darkness,
Eventually gliding down onto a grassy field
Just as the gas ran out.

From here I walked for tens and tens of miles,
Always taking the most difficult path
Whenever there was a choice,
For this would insure that I could end up
In some totally unvisited region
That was near impossible
Or hard-to-get-at in any way.

After hundreds of these
"Improbable" path choices
I suddenly came across acres
Of Lady's Slippers flowers;

These are very rare flowers
That usually appear in small bunches,
Growing only in conjunction with a rare fungus,
And, even, so, usually get picked—
But, there were millions of them.

After taking one last really difficult choice of a path,
I discovered entire fields of other flowers
Long thought to be extinct.
Some were Eve's Blossoms,
Which not been seen for thousands of years,
Historically valued for
Their life extending elixir,
As well as the original, lost,
Strain of Pearly Everlasting,
The flower that never dies,
And so I suspected that
I might be in virgin territory.

How would I know?
Well, for one, there were no paths left,
For even animals and their hunters
Had either long left or had never even been here.

Also, the flower colors were not like any
That I had ever seen before,
Not new colors, mind you, but, just, well,
Colors of different intensities and hues
That were not thought to exist in nature.
I saw true-blue roses, legendary no more.

I had chanced upon a land of strange rainbows
Of elfin-hued flowers: Red Delphiniums, Black Tulips,
Orange Fuchsias, White Marigolds, Bronze grass,
Yellow Violets, and even Adam's Apple,
Now growing from the ground!

Was this the original forest—the Garden of Eden?
Was I the first to return to this legendary paradise?
And then I knew that it was indeed the Garden,
For there, right in front of me,
Was a field of thousands of undisturbed
Golden nuggets on the forest floor.
Surely no one had ever been here,
At least not for a long, long time.

I reached up and put the apple back on the Tree.

X-MAS

A little guy and his friend, both from Jupiter,
Were investigating the potential of earthlings.

They were little, actually tiny,
Because the gravity on Jupiter
Would crush anything larger.

They were so happy to be able
To jump more than an inch on Earth
And so they leapt and bounded for a while,
But then got down to business.

"Look", said one,
"They have strings of lights
Outside their houses."

"Yes," said the other,
"But lights go on the inside,
Not the outside!"

"True, and... oh my Zeus!"

"What!"

"They cut down trees
And put them inside their houses."

"Trees belong on the outside!"

"Let's go.
There's no hope for these earthlings."

"Wait till Jove hears this one!"

THE UNIVERSAL ACID

As a boy in chemistry class,
I wondered, as did many,
About the following scenario often dreamt of:

I mixed two compounds, which, unfortunately,
Produced the ultimate acid.

Nothing could contain it.
It quickly ate though the container,
The floor of the laboratory,
And then even all the way through the earth,
Eventually sloshing some poor sap in China.

This, too is what happens to us, through education,
As our chemical-bio-electric nature is revealed to us,
Like some kind of giant shock,
After which we will never be the same again,
As perhaps some are now reeling from,
Well, maybe just a little bit.

The biochemical mush that is us,
When fully realized,
Leaves us stunned and astounded.
We grasp for what we once thought we were before,
But, it eludes us in the new light of learning.

The universal acid of such knowledge
Eats through all superstitions, folk tales, and myths.
Nothing can contain it.

We may come to even regret
Our learnings of this condition,
For it dissolves our container,
Leaving us floundering in the lurch.
It happened to me, too, beginning in fifth grade.

But, wait, it's not so bad, is it,
For what we are is what we are,
And we still have feelings, personality,
And more adventures of learning that await.

The light of education ever shines brightly,
Wherever it may lead.
Many dark alleys remain to be explored,
Given our new insight into the human condition.

EINSTEIN AS A NEAR TRAFFIC FATALITY

George Gamow told in his book,
'My World Line',
How he was conversing
With Albert Einstein
While walking through Princeton
In the 1940s.

Gamow casually mentioned
That one of his colleagues
[Pascual Jordan]
Had pointed out to him
That according to Einstein's equations
A star could be created out of nothing at all,

Because [at point zero]
Its negative gravitational energy
[mass defect]
Precisely cancels out
[is equal to]
Its positive mass energy
[rest mass].

"Einstein stopped in his tracks,"
Says Gamow,
"And, since we were crossing a street,
Several cars had to stop
To avoid running us down".

GOD ON TRIAL

"Jehovah's" trial for crimes
Against humanity begins thusly,
But ends well:

"Do you, God, swearest to tellest us the whole truth
And nothing but the truth, so helpest you God?"

"Which scriptures of what bible should I swear on?
There are so many."

"Oh; here's a Mormon bible
With a whole extra section
That was transcribed from
The golden plates You sent."

"I didn't send those plates."

"OK, let's not worry about that now;
We'll come back to it later.
You are truthful, are you not?"

"I can do no evil, and that includes not lying."

"Finally, a believable defendant.
What is your full name?"

"'God Damnit' is what I am usually called."

"Ha-ha, but what is your real and proper name?"

"None. I am what I am."

"Um, any aliases, like Lord,
Jehovah, Almighty, or such?"

"No."

"Are you sure?"

*"Yes, those are just some names
That people call me,
Plus even the very bad names."*

"But you do exist as you are?"

"Depends on what the meaning of 'exists' is."

"You know, like 'to be', being One that is."

"Depends on what the meaning of 'is' is."

"Is that your lawyer, Bill Clinton,
Sitting over there?"

"Yes, for he can get out of anything."

"But is he going to talk endlessly in your defense?"

"No, he has been going to 'On and on anon'."

"Good, now how come we can hear
You but we can't see You?"

"I am invisible, plus, you are schizophrenics."

"Hey, no name calling, order in the court!"

"I'll have a cheeseburger, no pickles, no onions."

"That's more like it.
So you mean we are just hearing voices?"

*"Yes—do you remember the study that showed
That 17% percent of priests are schizophrenic,
But only 1-2% of the general population is?"*

"Oh, yeah, but You're not getting off that easily."

"I am innocent."

"What did You do before You Created everything?"

"I was being made Myself by Myself."

"How did You do that?"

"Recursively."

"OK, anyway,
Did you have intercourse
with a teen-age virgin?"

*"Hell, no, she was underage;
I only date 30 billion year old women."*

"Still single?"

*"Yes, for as Mr. Always Right
I could just never find Miss Perfect."*

"So, Jesus was not Your son then?"

*"No, but he was a really good guy—
A human telling stories
That everyone expected to hear."*

"But, anyway, you are a 'He'?"

"So they usually say."

"Don't You know?"

*"No, for humans created Me in their own image
And with their own traits, so I am male."*

"Jealous of any of their other imaginary gods?"

*"I am above all that
Lowly human-type emotion stuff.
I am Perfectly Good
And absolutely totally full of Love."*

"Love is a human emotion."

"That is the only emotion I have,
For it is the ultimate one."

"So, You never do evil?"

"Depends on what 'evil' is."

"Well, as in things
Like harming others,
Except in self defense,
Stifling the growth of mind,
And creating false ways of living,
Arbitrarily, through use of imagination
Of what the concept of good 'should be'."

"I am not capable of evil. I detest evil.
I would hate Myself if I did evil.
It is unthinkable.
Then I would be in the category of a devil."

"Is there a Devil?"

"No, I would not tolerate any such thing,
For then it would sway humans to sin."

"You appear to be without fault,
But we still have to continue this trial."

"Thank you, but I have no-fault insurance."

"Did you murder almost everyone
On earth with a Great Flood?"

"Heck no, human nature is exactly
The way it is supposed it to be, as is.
What do you think!
God not a big fat goof,
That is, if He was involved.
He doesn't make mistakes."

"Some say that you invented the rainbow
To proclaim that You made a mistake,
Claiming that You would never do it again."

"Preposterous. Rainbows are an optical effect."

"Do You ever do anything wrong?"

"I can't. I am all Love."

"Did you give too much love, perhaps?"

*"Yes, I give near infinite amounts,
But there's nothing wrong with that."*

"What was the purpose
Of having dinosaurs around
For 650 million years,
Then extincting them via asteroids?"

*"Just playing around;
Actually, I had nothing to do with it."*

"What was the Intelligent Design in this?"

*"There wasn't any, for God dos not exist.
Can I go now?"*

"No, we know that nonexistence trick.
Whose side are you on in football games?"

"I don't take sides or play favorites."

"Then where do humans
Get all these ideas about You?"

"You know humans—they just make things up."

"Is there a Hell,
Like maybe in the heart of the sun?"

"No, there is no Hell.
I wouldn't torture my beloved creatures
If I were God.
Would you torture a kitten?"

"Some would, but, hey,
It is You that is on trial here, not us.
We only have our human nature
That You may have given us
And it can often go astray."

"True, plus I am a nice Guy, the nicest ever.
I would not fill your cup
To the brim with temptations
And then expect you not to spill it;
I'm a giver, not a taker.
Pure love is all giving;
There are no strings attached."

"Thanks.
Does our free will have to match your will"?

"Heavens no,
For that wouldn't be free will, would it?"

"So, there's not even a purgatory,
Like somewhere on Venus?"

"Negative."

"How do humans come up with all these things?
They make You out to be some kind
Of strict enforcer father figure type."

"That's it; they modeled the family experience."

"Is there a Heaven?"

"Yes."

"Ah-ha, where is it?"

"On Earth.
What more could human beings want?"

"Oh, well they want everything
And even think they are
Special and above all else,
Some even above their own kind."

"Nope, humans are as organic
As anything in nature.
Anyone can see that."

"Well, we have imagination."

"Yes, a gift of Nature, but that's all it is."

"Did You publish a book?"

"Yes, but no, for ghost writers wrote one."

"Any movies coming out?"

"No, it would be hard to beat 'The Dark Knight'."

"Were Commandments were ever issued?"

"Love does not command; it frees."

"That's true.
So You are innocent of all charges
And plead not guilty?"

"How many times do I have to tell you.
I am Absolute Good."

"Ever tell a white lie?"

"No way, Jose. I am the Truth."

"Ever peek at a naked person."

*"Of course, people are made that way.
If He didn't want it that way,
They'd be born with clothes or fur.
Some fools even put fig leaves
Over Eden's artwork."*

"I must confess to You, God,
That I sometimes think of people naked."

*"No sweat, plus that's also a way
To make public speaking easier.
I am naked Myself. It's OK."*

"Ever stick gum somewhere
When no one was looking?"

"No, for I was looking."

"You are a saint!"

*"Higher than that. I am Perfect,
At least before I got conceited about it."*

"Ah-ha."

"Just joking."

"Did You make Cosmic Jokes,
Like, in sexual human anatomy,
Putting a toxic waste dump
Near a recreation area?"

"God does have a sense of humor."

"How come You didn't give humans everything?"

*"If I gave them everything,
They'd have no place to put it all."*

"A dictionary has 'everything'."

"In a way, plus Wikipedia is good, too."

"How come birth certificates
Have expiration dates,
Some even sooner than later?"

*"They must, otherwise,
Evolution wouldn't work."*

"Did some monkey types
Descend from the trees?"

"Yes, for your DNA matches theirs 98%."

"So, evolution is true,
But not you as a Creator?"

*"I keep telling you,
Leaving signs all over the earth,
You fossil to be."*

"You don't rule or lord Yourself over anyone?"

"Love serves; love does not rule."

"We have witnesses to some of your crimes."

*"No one can witness Me,
Besides, they made all that up."*

"Likely story.
Did you choose a tribe and tell Moses
To crush some other tribes?"

"Those are just ancient Jewish legends."

"How come Moses didn't ask for directions
When he was lost for 40 years in the desert."

"He's a man; they never ask."

"Ever let someone just make it
Through a developing traffic accident?"

*"What! And let some other poor sap
Get hurt or die instead?
You don't know Me very well."*

"So, You don't write scripts
For our human soap operas."

"No, for truth is stranger than fiction."

"Why are You invisible?"

"I am a figment. Have faith."

"What's faith?"

"Belief in the invisible unseen unknown."

"You can't get off the hook that easily.
We can still try you in absentia."

"I'm being very cooperative."

"Thanks. Now, Mr. God, Sir,
Did you send a plague of locusts
To harm the welfare of humankind?"

*"I wouldn't think of it;
Harmful options don't even surface
In my mind for consideration."*

"No lightning bolts?"

"That was Mother Nature, not me."

"Well, as you are a self-made Man,
Then what stuff did You use
To make Yourself out of,
Plus all that is?"

"I didn't make all that is;
I only made Myself out of
The fundament stuff available;
Then I accidentally made humankind
From the same stuff,
Some debris that I threw out."

"So, you are not at all responsible
For Mother Nature's doings?"

"No, nor did I make the universe,
For I am made of it."

"You are not fundamental and absolute?"

"No, for a system of mind and emotion
Like Mine or yours requires moving parts.
I am perfect, however."

"That's still a lofty position."

"I am just fortunate to be as I am;
I never look down on anyone less;
My talent is a given;
I can't even really take any credit.
I am just further along
In evolution than you are.
Cats, too, have reached a kind
Of perfection for their form."

"You evolved beyond the material plane?"

"Yes, I am pure waves and fields
And thus not seeable.
You all will get there someday, too.
I just helped you all along the path,
With only your best interests at heart."

"We will all evolve to become Gods, eventually?"

"Certainly."

"You don't interfere in our world on Earth?"

"No, for then you would miss all the fun.
Knowing everything is not really that great."

"There would be no surprises."

"Exactly."

"Do you overrule all or part of reality in any way?"

"No, I'm not bossy."

"Do you underlie all or part of reality in any way?"

"Nope, as I said, I am in this universe
And therefore of this universe;
I am just higher up the food chain."

"So, in our terms,
You are just a very powerful but loving alien."

"That I am.
And if any hostile Ones approach me,
I will defend Myself."

"Thanks, for that may help us, too."

"True, but you are all completely free to be and do."

"How come You allow/give this to us?"

"It's the greatest gift that Love can give."

"Thanks, again.
You seem a good Guy,
But we still have a few more questions,
Plus, you know, we can't really consider any gifts
That You gave to us when we make our ruling;
I hope you understand,
For we are often approached with bribes."

"Money talks."

"For me it just usually says 'Goodbye'."

*"But when it returns you might say,
'Hey, glad to see you; I've missed you;
Where have you been all my life?"*

"You're a fun Guy.
So, what is all this holy-holy admiration stuff
That humans do in and for your Name?"

"I don't know; it's really weird, isn't it?"

"I thought You knew everything."

*"Well, by staying out of the way,
I chose not to know."*

"What made the stuff
That we and You are made of?"

*"I'm not sure;
I only know everything
From Me onwards;
That stuff could have
Appeared in the universe
From somewhere else,
Or have been here forever,
Or appeared via some kind of possibility;
It is not marked as holy or unholy."*

"Well, that's immaterial, anyway.
Back to our probe."

"I ain't never did anything terrible nohow!"

"Ever do anything wrong at all?"

"I threw some litter into space
Because there was no where else to put it."

"What litter?"

"An excess atoms that then made your world."

"Well, no harm done."

"Thanks."

"Do angels exist, having wings and all that?"

"No, not as humans have defined them.
Wings are useless in space; there is no air.
There are more ETs than Me, however."

"We thought so.
Is there a Bigfoot?"

"Ha, ha. Those are just hoaxes put forth
By some hicks in the southern US."

"Isn't 'hick' a bad name?"

"No, I am just describing an actual fact,
For which the word 'hick' is perfectly descriptive.
I have to use words that you can understand."

"So, You've never been seen,
And just about everything bad
That was said about You
By humans is false; so, what's left?"

"Not much, just Me as not 'God'."

"But You created us; you helped us along."

"Well, in a way, but that was quite inadvertent.
You would have formed
Somewhere sometime anyway.
Some of my 'trash' formed
Your solar system;
Then you evolved.

Your population was down
To less than a thousand once,
And I guess some of my
Good vibrations rubbed off on them
As I passed by on my way
To pick up some rare elements on Pluto.
I was building a new house
That can withstand all eternity.
The weather in space is always bad;
It's full of radiation of all sorts."

"Strange weather all over the earth, too."

"There are many hurricanes that began
From a hint of a wisp of a breeze."

"Mr. ET, is there way to tell
The future of the weather?"

"The 2010 Farmer's Almanac just came out."

"So, how do we speed up evolution?"

"Takes time,
But you could enhance
Your own chemistry,
As I did."

"Sounds dangerous."

"It is; I was a Jeckyl and Hyde for a while."

"Ah-ha, that's when
You committed crimes against humanity!"

"No, I was far away,
Plus that was 35 billion years ago."

"Oh, but do You have an alibi?"

"No, I was all there was then,
But I have pictures."

"Let's see."

"I don't have then with me,
But they are very similar to those
Taken by the Hubble telescope."

"You were there among those
Trillions of stars and galaxies?"

"Yes, but I was already semitransparent by then."

"It would be like one of those
'Where's Waldo' puzzles."

"You'll just have to take my Word
If you cannot prove otherwise."

"What is the purpose of life?"

"To live."

"What is life?"

"You are life."

"Is life and all really just a bunch of
Atomic spinning things
Of various compositions?"

"That's it."

"Nothing more?"

"There can be no more, for that is all there is."

"Why do we keep hoping for more?"

"Greed and having no gratitude, but,
Still, you are a sparkling billion year product,
And quite amazing."

"We are pretty cool when you think about it."

"That's all it takes to appreciate life."

"Any other universes?"

"Sure, but many did not amount to anything.
However, I am going on vacation
To a good one next week."

"Be sure to send a post card saying
'Wish you were here',
That is, if there is oxygen there."

"Will do.
Lucky for you here that bacteria and plants
Came about and made oxygen.
Thenceforth you began as you."

"Yes, a lucky break;
Oxygen was a mere waste product
From photosynthesis."

"See, all is as it seems.
No need to invent any supernatural Intent
To blame or thank for anything."

"All is as it did?"

"Yes, that's why it took so long."

"Indeed, A true God type Creator
Could have done it instantly,
Not even needing 6 days,
Or getting tired on the 7th."

"Yes, but the All is an origin, not a Creator.
The ground-state was always around,
And so there was no creation, and no Creator."

"Yikes, then what should we do?"

"Just be."

"OK, good advice, but,
If we ever find that there was a Culprit Creator
Who committed some of the very crimes
That His commandments spoke against,
Like murder, destruction, or hatred,
Then He is really going to be toast."

"As He should be,
For those acts would
Have been unconscionable,
Especially for Someone
Of that high stature."

"Thanks for your testimony.
We'll call it the Third Testament.
Your judgment day is near at hand.
I'm calling a one hour recess."

...

"All please rise."

"The court finds You not guilty on all counts,
Due to lack of evidence, plus Your good nature."

"Evidence for those like Me
Is not even conceivable."

"True. Thank you everyone.
Please bring in the next case."

Austin walks in.
"Did you leave the toilet seat up in a household
Where there were females present?"

"Well, maybe, yes I did, but..."

"100 years of hard labor in Siberia."

THE TIME CAPSULE

Since one million years had just passed by,
They, of the future, prepared to open, nigh,
The absolutely sealed container's prize,
Of a capsule made so carefully that it did survive
Without damage, being totally impregnable
To any outside influence imaginable.

They expected to see, perhaps, some old relic,
But certainly nothing alive that could tell of it,
For it would be hard to imagine, even then,
That some organism could keep on going its ken
Over its course of a million years later,
Sealed inside this tight container,
Unable even to exchange energy's spark,
This metabolism being the hallmark
Of life and all that quacked or quarked...

And, so, they did not at all expect something
In there that would be flapping its wings,
Gasping for air, or anything at all of life's song,
Wondering what had taken so long.

Well, they were right and they were wrong,
For in the time capsule that was planted so long,
Several things had with it come along...

One was a plaque, of numbers low and high,
And containing some primes and pi,
Another, some essays of the future—
Some, like Austin's, quite mature,
Along with maps and other items of the world
From those times when the oceans swirled;
But, the last, one perhaps not intended,
Was a microbe—an extremophile—
Laying there quite contented all the while!

Well, they soon laughed, loud and long,
For they were in between right and wrong
About what could survive from so long ago,
For, it was really walking mighty slow!

AFTER THE STARS HAVE GONE—
THE FINAL, SILENT DARK

THE LAST CHANCE SALOON (CASINO)

Entropy is always the winner in the end,
When there's no more money left to lend;
Meanwhile we stabilize, in nature's way,
Rearranging resources temporarily.

Prelude

Going beyond our very old obsession, so vast,
Of how it all began, back in the distant past,
But, retaining our search for meaning, from that,
We nw turn to how will it all end, this and that,
Whether becoming collapsed, expended, or flat.

Is there is some deep meaning in all that?

Yes, for it is there in that future distance,
We'll find, or not, the end of our persistence—
Whether or not we are at all forever resistant;

Whether all that was, and what was did and done
Will be of any long-lasting benefit to anyone,
Of what destiny awaits, if there ever was one.

Endings are important to us, for what we're about,
Because we believe that how things turn out
Implies what the beginnings ultimately meant,
Of what, or not, is our place in the firmament.

As an ambitious species of nurture and nature,
We are now and always pointed toward the future,
For, of the three forms of the chimpanzee:
The common chimp, the bonobo, and us, we
Are the only chimp who went beyond the trees...

And, more importantly, even out of Africa, freed,
By that exodus, which laid down, indeed,
From that experience, the urge and the need
To move on, exploring, ever planting another seed.

The horizons on Earth sufficed us, as in "time",
For many millennia, but now the horizons' climes
Are broadened, through cosmology and physics,
And so they can well inform us of our prospects.

The future matters to us, for very basic reasons:
We wish to offset our mortality, our pleasin's,
To know if humanity's works, for every season,
Will be remembered, or lost; for nothing, even.

The Final, Silent Dark Marches On...

Time hurls a million waves of is displacement
At us, yet we are still there—our replacements:

Time, ever gray with age, hurls its changes, then,
'Gainst existence's rock, time and time again,
The entropic seas denuding the sands,
Yet, energy is preserved via science's wands.

Reminiscence weathered, but could ne'er wither,
For, in those mists of time; yesteryear yet appeared.

Would the prospect of a "Big Crunch" bring on phobia,
Such as an ever more confining claustrophobia?

Seems a better thought, somehow, though no picnic,
But more pleasing, if the universe(s) were also cyclic,
Although then all would still be really crushed
And forever lost, gone headlong into the rush.

We expect cycles, for all the days and seasons
Embedded this in our ancestors, into our reasons,
Since, at least, the periodic supplies some rhythm,
A pattern—the rolling hills of lives onward driven.

As for the cyclic, endless repetitions, they, too,
Would seem to revolt more of us than just a few;
As, too, perhaps, would some infinite abyss of time,
Which, too, grants us neither reason nor rhyme.

Does the drama go on forever, or does it end?
What do the visions of the future portend?

Doesn't it all have some purpose meant—
A goodly end of all of it to us might it present?

Is our higher mammal time, certainly,
But of such short parentheses within eternity?

It's only a finite time, then, which, too, tends
To horrify many, and more, as the universe ends,
Such as told by Robert Frost, a name of chill:
In heat or in cold, known as fire or ice, still.

Should we not believe in God since nothing lasts?
Well, if nothing lasts, then of what our purpose past?

Is a purpose really required, so constructive,
Or would that be really quite restrictive?

No realm could really be special or sent,
Its becoming being of some specific intent,
For, all arrived here of causeless accident.

Is there anything wrong with the freedom to be,
Anywhere, any how, or any time during eternity?

No.

Should we rail against the law of entropy,
The "heat death" of thermodynamic energy,
The second of its final laws, we see,
Because it would destroy all of history?

Well, there are so many ways for disorder to be
Than any one ordered state specifically.

Would even a heaven on earth become a misery
If it, as it might, contain no more novelty?

Must there be an end to our revelry?
Can't we, at least, hibernate eternally?
Won't all matter, too, last eternally?

Will Shakespeare's works live on, paternally?
Is this not a Wagnerian struggle for eternity?

**Science can settle whether a Last Day
Is ever going to come this way.**

Only a decade or so ago, with some consternation,
We discovered the universe's large acceleration,
This expansion even increasing, onto some thin disaster,
The galaxies getting further away, ever and ever faster...
Then, one last snapshot taken, for all to remember.

The accelerating expansion of the universe's rafters
Means that the universe will cool even ever faster,

So, any conceivable forms of the future's life prolongers
Will have to keep themselves ever more cooler,
Think more slowly, and hibernate ever-longer.

One day the protons will fade away,
Leaving but dark matter, electrons, and positrons.

*The Waves of the Ancient Swells
Of Time's Forgetting Tides
Swept Ever On...*

*As Time, now hoary with age,
Hurled forth its ashen change,
The charge ever san, pale and colorless,
That force born to summon decay, so endless,
'Gainst Nature's Universe each and every day;*

Time and time again, Time fed all upon,
In its bloodless, white and waxen way;

But, this everlasting rose would not fade,
Its luster even brightening by the day,
Ever unsuccumbing to the sickly, peakèd
State draining drawn the life away.

Entropic seas yet denude the mountains,
Yet, this enduring flower, never-endingly
Has cast Deathly Time aside, for now,
Ceaselessly somehow thriving on,
To that which was the near imperishable,
The flame of beauty still inextinguishable,
Forever celebrated as immutable,
Gaining its seemingly perpetual permanence
From the undying love of the glorious truth.

Yet, everything was moving apart, cooling off,
The big slowdown not really so very far off;
Ultimately, even the black holes of late
And the lightless planets would dissipate.

The primordial soup, once so rich and hearty
Was now a thin gruel that couldn't serve the party.

One day, every particle would be moving away
From every other particle, so much out the way
That they won't even be able to see one another;
Thus, for all intents, motion will have ceased forever.

Our spurt of life, followed by an infinite stretch
Of dark equilibrium, was but the briefest sketch—
A warm and fuzzy stage, so interestingly active,
Whose time, relatively, was but infinitesimive.

Yet, we were there, in all our glory,
For whenever else could we be?

In the future, uncounted societies of
Overlapping minds accumulate, with love,

In island redoubts, their preserved data burning
With a vital remembrance, in which, returning,
Past is the present and future, they all reliving
The data, even animating it and ever altering.

Without any new enrichments, the present and future
Reprise the past, in this retreat from external nature.

Their candles would have been nearly invisible to us,
They enduring, by diminishing, so as not to exhaust.

They made few new memories, a kind of blind sight,
For whatever realities had ever existed out of sight
Of their own mental structures were now fractured,
And thus not much different from those manufactured.

The Penultimate Part of the Final Dark

AN ESCALATING ONE WAY TRIP
FROM A FLUKE TO OBLIVION

The majority of the energy
Of the universe is dark today,
Although everything else passes
Through it in every way.

It's everywhere,
Having a component
That repels its own state,
Which cause the expansion of
The universe to much accelerate.

DARK ENERGY MATTERS: THE ESCALATION

We're on a one way trip from the quantum fluke,
That maximal energy within old Planck's nook—
Heading toward the oblivion of sparse expansion,
All that we ever loved and knew going to extinction.

We sent message of early warnings to some,
In those castles of illusion, yes, many a one,
That they would face the decay, not so far away,
Of the heavy particles, the "proton pause", one day.

No self-assembled granularity can endure
Forever, but must return to the substructure,
And, so, the lives must all transition, it seems,
From heavier to much lighter regimes...

Although this, too, would not be permanent,
All destined to be swallowed by the firmament.

We have often asked why some space exists,
Why it permits the countless to briefly persist
On Mother Earth nourished under Father Sky—
All of those finite sparks that light and die.

There were those who endlessly debated,
Whether to live in their virtuals unabated,
Or press forwards and outwards, of delirium,
To seek new localities in the mysterium;

But, the pauses of the heavy particles continued,
And so there was nowhere to go for the retinued.

It was much simpler once, in those days of old,
When we thought that universes didn't go cold,
But that they expanded and collapsed,
Still destroying all, yet ever giving more to last.

And, well before that, once upon a storied time,
We simply made it all up, with tales and rhyme,
In place of any physical observations,
Or of all our revealing experimentations.

...

The past was now a reef of dead accumulations,
A graveyard of various useless informations,
Which, despite their splendorous beauty,
Could not provide a novel futurity.

...

The last one of us, born of the sparkness,
Kept a window to the outer darkness...

S/he looked out, from a once brightly
Colored and sparkling inner reality,
Into the dark abyss...

There was nothing out there,
All being so lonely and bare—
No more singing of life's song;
For now everything was gone.

The Final Epilog

MULTIPLE VERSES

Our fruits are of a universal seed,
Are yet another yield of All possibility treed,
For siblings elsewhere in the entropic sea
Are also born of such probability.

There could not have been any special time,
One that was privileged over any other chime,
Nor any special place, nor any specific form
Arising out of the necessarily causeless realm.

Even those locally specific dates and places past
Of the events' novel memoirs could not ever last,
They being writ on water, with no meaning vast,
Disappearing in significance so very fast,
Since it's only the universals that last.

...

The protons were all gone from the show,
Having decayed so very long ago,
Into positrons—ever canceling the electrons,
But emitting the fleeing light of photons;
There being, of course, an equal amount
Of protons and electrons in the count;
And, of course, along with all the protons,
Went all of the atomic elements, the end,
All of their forms becoming myth and legend—
As they were still dreamt in night dreams,
Those forms that we once had, so it seemed.

S/he, as many of a luckily adaptable kind,
Had long since lightened and lighted the mind
With the dwindling electrons, and precious photons—
That beginning light of ancient times, growing wan.

Ours had been the only line in the uni-verse,
One that had become sentient, with proto-man first,
The rest of the cosmos being but a colossal waste,
A foreboding, harsh, and very dangerous place.

S/he was now the only one left,
Having outlived all of the rest.

...

The universe was near crumbling away,
Having run out of space, time, and all its sway.

S/he was dispersing, melting, into the vacuum, lone,
But, s/he held on for another thousand years, alone;
And, then, s/he, too, was gone,
Being the last of the hominid's song,
Of all that was sapient: the *Magnificat*,
The composition of Earth's sweet plot,
The greatest symphony that was ever sown,
Now having faded into the unknown.

From near nothingness our forms became,
And into the same must go our remains.

If the unknown be such, 'though it's otherwise;
But, still, if it's yet called unknown, then the reply
Is still, for sure, that we're free to be, anywise.

If you've shed a tear, reading here,
For both the far, and the near and dear,
It won't make their graves green again;
But, it's possible that life could begin again...

Be of Good Cheer--the sullen Month will die,
And a young Moon requite us by and by:
Look how the Old one meagre, bent, and wan
With Age and Fast, is fainting from the Sky!

(A Fitzgerald quatrain that's not in his Rubaiyat)

THE ETERNAL RETURN

Behind the Veil, being that which ev'r thrives,
The Eternal Multi-Cycle has ever been alive.

Some time it needed to learn Everything for,
And now well knows how these bubbles to pour,
Of existence in some meant universe,
Those that wrote your poem and mine, every verse.

So, as thus, thou lives on yester's credit line,
In nowhere's midst—now in this life of thine,
As of its bowl our cup of brew was mixed
Into this state of being that's called "mine".

Yet worry you that this Cosmos is the last,
That the likes of us will become the past,
Space wondering whither whence we went
After the last of us her life has spent?

The Eternal Saki has thus formed
Trillions of baubles like ours, and will form,
Forevermore—the comings and passings
Of which it ever emits to immerse
In those universal bubbles blown and burst.

So, fear not that a debit close your
Account and mine, knowing the like no more;
The Eternal Cycle from its pot has pour'd
Zillions of bubbles like ours, and will pour.

When You and I behind the cloak are past,
But the long while the next universe shall last,
Which of one's approach and departure it grasps
As might the sea's self heed a pebble-cast.

THE MATERIAL WORLD

Our so-called "material"
Is but the secondary,
And, thus, non-elemental,
As emissions
Of opposite particle pairs.

All this, such as its other forms
Of positive mass-energy
And negative gravity energy,
All sum and cancel to a zero, literally,
Being but the arbitrary and fleeting phantoms
Of the temporary particulars of specifics
Sprayed from the uncaused, eternal,
Fundamental, and necessarily
Indefinite ground-state beneath
That could have no real intent
Or direction to it;

For there could be no "before",
Nor of Nothing, nor of causes
Beneath causes of infinite regress.

A complete freedom is the glorious result,
Within our form, of course,
With no strings attached,
Much better than a constraining "purpose",
Yes, when you really think about it.

THE NEAR FUTURE

Life should be euphoric, like spring fever,
As in those rare moments of ecstasy
When one is in the zone and cannot miss;
So—let all aversive substrates be removed!

The higher modes of being that await
The future-chemically-enhanced
Will make today's primitive mind-states
Seem as a child's tin flute to a symphony!

Mind reaches out to see what's possible
And what's not, like particles forming
In the quantum world, but, better than that—
Mind makes the impossible possible.

Mind is the ultimate of all there is
It is the universe—billions of years
Of primordial material—complex.
So, what more could human beings want?

The secrets of the universe are all found—
All exists out of consciousness, the ground.
Blame, soul, free will, and God have all fallen—
But it will take a thousand years to sink in!

— THE END —

www.ingramcontent.com/pod-product-compliance
Lightning Source LLC
Chambersburg PA
CBHW071407170526
45165CB00001B/202